TOXIC EXPOSURE

TOXIC EXPOSURE

THE TRUE STORY BEHIND
THE MONSANTO TRIALS AND
THE SEARCH FOR JUSTICE

Chadi Nabhan, MD, MBA

JOHNS HOPKINS UNIVERSITY PRESS | BALTIMORE

© 2023 Johns Hopkins University Press
All rights reserved. Published 2023
Printed in the United States of America on acid-free paper
9 8 7 6 5 4 3 2 1

Johns Hopkins University Press
2715 North Charles Street
Baltimore, Maryland 21218-4363
www.press.jhu.edu

Library of Congress Cataloging-in-Publication Data is available.

ISBN 978-1-4214-4535-9 (hardcover)
ISBN 978-1-4214-4536-6 (ebook)

A catalog record for this book is available from the British Library.

Special discounts are available for bulk purchases of this book. For more information, please contact Special Sales at specialsales@jh.edu.

CONTENTS

AUTHOR'S NOTE

The incidents and facts shared in this book are all based on true events and on what I have experienced as a medical oncology expert witness in the Roundup litigation.

As of the date of this writing, there remain additional ongoing lawsuits against Monsanto, the manufacturer of Roundup, which was acquired by Bayer in 2018. The book describes how I got involved in this litigation in 2016 and what transpired in the first three cases against Monsanto, brought by four different plaintiffs, that went before a jury. The book does not address subsequent trials after the Pilliod trial concluded in 2019.

The facts described herein are based on and derived from publicly available information, documents, pictures, testimonies, and/or statements available in public court filings, trial transcripts and exhibits, regulatory agencies' websites and data, and other public dockets or repositories, publications, media statements, press releases, and/or other means by which information is readily and publicly available for all consumers. All the information and figures shared in this book are available through public records and posted on various law firm websites for anyone to see, read, and post elsewhere. No information in this book is subject to any court-approved confidentiality or protective order restrictions. Some of the trial transcripts were slightly edited for flow and consistency, but the subject and meaning were never altered.

No names or characters have been changed. The book comes out after several jury trials have been concluded and after judgments have already been rendered.

Certain dialogue and events are reconstructed to the best of my ability, and just for flow and easy readability. Every effort was made to avoid revealing or disclosing any attorney-client advice, confidences, or work products, and without waiving in any way any such privilege, protection, or other legally recognized or otherwise available privileges or protections for any character in this book.

All opinions and dialogue cited here—outside of the quoted depositions and/or trial transcripts—are my own opinions, ideas, feelings, emotions, statements, and recollection of events, including the personal details I share, recounted to the best of my ability.

I just couldn't seem to escape it.

Nearly four years after the trials, more than six years after I'd been unexpectedly dragged into what would become one of the largest product liability lawsuits in American history, I still seemed unable to shake myself free.

Even here in my gym—the One-on-One Personal Training Center, tucked into a nondescript office park in Northbrook, Illinois—I couldn't get away. A commercial playing on the gym's TV reminded me of everything that had overtaken my life. The events of the past several years had put almost unbearable stress on my career and my family, nearly driven me to despair a few times, and certainly contributed to a feeling of physical burnout and mental exhaustion.

The words appeared on the TV screen in front of the treadmills as I dutifully plodded along on my warm-up. If someone was animating my life story, they might draw those words floating out of the TV and circling my head as I tried to brush them away. Those words have defined my life for more than half a decade now and will probably continue to do so.

Monsanto.
Roundup.
Lymphoma.
Settlement.
Dollars.
Billions.

"If you or a loved one was diagnosed with non-Hodgkin lymphoma or leukemia after using Roundup weed killer, pay attention to this important message," blared the announcer. The screen showed a smiling, gray-haired man in a golf shirt, someone who looked like he could be your grandfather, blithely spraying Roundup on the weeds around his tomato plants. A screen crawl of erratically capitalized and exclamation-point-infested text added to the sense of urgency: "A ten-billion-dollar Roundup lawsuit settlement was recently announced!!! Time is running out!!! Get a free Roundup Lawyer Case Review before Your Right to Sue Is Lost!!!"

A 1-800 number for a law firm then appeared, flashing in different colors. "Roundup Lymphoma Victims," the commercial concluded. "We Are Here for You."

I shook my head. "This is draining," I said to Sasha, my trainer, as he arrived for our 6:30 a.m. appointment.

Sasha looked up at the screen and then back at me. He smiled. "Well, Chadi, I know how exhausting this was for you," he said. "But you did help some people, didn't you?"

We did.

Justice *was* done.

I strongly believe some innocent people who were misled by a corporation too interested in preserving its image and profits got the compensation they deserved—and the corporation, in my opinion (and, more to the point, in the judge's and jury's opinion) got what *it* deserved.

It took three trials, countless depositions, and arguments that pitted some of the highest-paid corporate defense attorneys in the country against a team of determined and committed law firms and lawyers who were willing to fight for the people who claimed to have suffered harm from Roundup. At several points along the way, the cases were on the front page of just about every major newspaper, magazine, and news site in the country.

The proceedings were closely watched by millions, and their outcomes have shaken the foundations of corporate America.

But Sasha was right. The outcome was a victory for the little guy. Granted, the little guy had to share some—maybe a lot—of the $11-billion-plus award with the lawyers. Still, the approximately 100,000-plus plaintiffs who received some compensation—people who suffered from a cancer that I believe was caused in some by using what they'd had every reason to think was a harmless home-gardening product—certainly got a measure of justice from what were, in some sense, the trials of the century.

And I was right in the thick of it.

I'm Dr. Chadi Nabhan. A hematologist and medical oncologist with almost twenty years of experience in treating non-Hodgkin lymphoma, I was a medical oncology witness for the plaintiffs in the now-famous Roundup trials. The trials I'll discuss in this book have been concluded, but in some ways legal action is still ongoing, as I am reminded whenever I see a commercial aimed at those allegedly harmed by Roundup, or run across another related news story, or hear about a new case being filed. I was among many witnesses who committed hours, days, weeks, months, even years of their lives in an effort to tell the truth and expose the facts of that Monsanto product. One goal unified us: defending patients allegedly harmed by Roundup.

I'm also Chadi Nabhan, father, husband, son . . . and immigrant. I was born in Syria, and although I arrived in this country thirty years ago and consider myself a proud and loyal citizen of my adopted country, I, like many first-generation Americans, feel a continual need to prove my worth. And while I embrace the positive, can-do spirit of America, I haven't lost all of the cynicism—some would call it realism—that comes with growing up in a part of the world where humans have lived and interacted for 700,000 years (it's true: modern Syria was founded in 1946, but the civilization that exists there goes back to antiquity).

I know what's it like to see people treated unfairly, dismissed, swept under the rug. And I don't like it.

When I was asked to help, to use my training and experience with this disease to help deliver some small measure of justice to people who had done nothing to deserve the misery that had befallen them, I said yes.

I'm still paying the price, in some ways. Granted, it's a much lower price than that paid by the patients who were allegedly harmed by Monsanto. Yet, on balance, it was worth it for me. Every stressful flight to San Francisco, where the trials were held; every long night prepping with the lawyers; every moment being interrogated and baited by the company's attorneys; every anguished discussion with my wife and family, addressing their concerns about how much time and emotional capital I was investing in this—I have no doubt it was all worth the sacrifice.

It confirmed to me that greed and arrogance are rampant at the highest corporate levels. It also showed me that sometimes that kind of behavior can be exposed, and that the right people, the honorable people, can win in the end.

"This verdict is, without question, truly historic," said attorney Brent Wisner, who represented one of the plaintiffs. "This verdict is groundbreaking. It's precedential."[1]

And it was also controversial. "Did a jury ignore science when it hit Monsanto with a $2-billion verdict?" was the question raised in a *Los Angeles Times* column on one of the trials I testified at.[2]

I hope the jurors didn't ignore the science. Because that would mean I didn't do my job, which was to show the jury that Roundup (and its active ingredient, glyphosate) could cause cancer, specifically non-Hodgkin lymphoma, in some situations. In May 2021, one of the federal judges who played a prominent role in these trials, the Honorable Vince Chhabria, paid me and other plaintiff experts a compliment in denying one of Monsanto's motions, saying in his ruling that Monsanto "has lost the 'battle of the experts' in three trials."[3] I felt pretty good about that, especially after the way Monsanto had attempted to exclude my testimony, challenge my opinions, and dismiss the opinions of the plaintiffs' other experts.

As draining and demanding as the experience was, I must admit that it was also dramatic. In many ways, the experience was akin to being a character in a John Grisham novel or a modern-day version of *Silkwood*.

I'd like to tell you the tale—not as a legal expert, a journalist, or an environmental scientist, but from my ringside seat as one of the medical oncology witnesses. I was someone who had never testified in court as an expert witness before but suddenly found himself testifying in three highly visible trials within one year.

I invite you to see the American judicial process the way I saw it, at the highest levels, behind the scenes, and with life, death, and billions of dollars in the balance. Reading this book, you can sit behind me and watch epic trials. You'll see a quintessential David-versus-Goliath battle unfold before your eyes.

So, let's turn down the volume on those commercials from law firms still seeking to wet their beaks in the deep and muddy waters of this far-reaching and dramatic case. I'll tell you the *real* story . . .

TOXIC EXPOSURE

1

The Phone Call

Thirty-five thousand feet above the ground, I sat anxiously on United Airlines flight 222 heading to San Francisco, watching the vast tableau of the American West unfold outside the window of the 757.

I turned away from the window and sipped my coffee. It was July 2018, and my task at hand felt as monumental as the valleys and mountains below: reviewing what seemed like an endless stream of documents, depositions, articles, and scientific literature. For once the balkiness of the in-flight Wi-Fi came in handy, as the malfunctioning system meant I couldn't constantly check my messages; instead, I focused on the materials in front of me, preparing for what some were already calling the trial of the decade.

I'm not a big fan of flying. I recall that I used to enjoy turbulence when I was growing up; it felt like a roller coaster or as if I was in some 1980s computer game, like Lode Runner or Maniac Mansion. That changed as I grew older, and my fear of flying certainly intensified when I became a father to twin boys. But I had to suck it up, since my job as an oncologist, researcher, and administrator mandated frequent travel. During this trip, a few air pockets provided just enough turbulence to jolt me out of my concentration. I decided I'd read enough and looked back out the window. The arid landscape had yielded to the verdant mountains of Northern California—mountains that, at another time of year, would be engulfed in smoke from the catastrophic wildfires that have plagued this part of the state. Truth be told, I could feel flames of nervousness licking at my insides: I was anxious about the prospect of getting up in front of a judge and jury, not to mention skilled attorneys for one of the world's largest agrochemical corporations. But I had vital information to share with them, information about a common household product that, I now firmly believed, was in some patients causing non-Hodgkin lymphoma, a form of cancer. Despite my anxiety, I was eager to share this information with the public. I was hoping that our justice system would come through and acknowledge the validity of the evidence. At the same time, I knew this trial would not be easy, as dozens of attorneys were suiting up to defend Monsanto day and night.

This trip to testify in such a high-profile case was the culmination of a two-year-plus journey that had started in the spring of 2016 when I was contacted by the Miller Firm, a Virginia-based law practice. The firm had called to inquire whether I would be willing to serve as an expert witness in litigation against Monsanto, an agricultural behemoth. The Miller Firm represented several plaintiffs who claimed they had developed non-Hodgkin lymphoma as a result of their use of Monsanto's herbicide Roundup. It wasn't uncommon for me to get calls like this, but I had never testified in court as an expert witness. The calls I got were usually related to medical malpractice lawsuits, in which I would be asked to review patients' medical records to determine whether an oncologist had deviated from the expected standard of care. I've always had an uneasy feeling about being engaged in such cases, believing that many are won or lost based on the theatricality of the lawyers litigating the case as opposed to true facts and science. This belief deterred me from agreeing to be an expert witness in most cases. Add to this that my workload has always threatened to surpass the hours and mental bandwidth I had available, and you'll see why I'd always found an excuse to politely decline.

Something about this call was different, though. I had received an email from the firm a week earlier, saying that one of the physicians they had worked with on a bladder cancer litigation case had recommended me. I knew that physician very well; we were friends. So, I had agreed to the phone call (as opposed to politely declining via email) as a courtesy to him.

The call started with two enthusiastic lawyers on the line, who identified themselves as Tim Litzenburg and Jeff Travers. Unlike attorneys who in the past had pressed me, they were pleasant and friendly.

"Dr. Nabhan, thanks for taking our call," said Tim. "I know you're busy, so I'll get right to the point. Have you ever heard of Roundup? It's the most commonly used weed killer in America."

I recalled seeing containers of it on the shelves of various stores near my home in Deerfield, Illinois. I had a vague memory of a bold logo with black and green letters sitting in a kind of swoosh.

"Sure, " I said, "I've heard of it."

"Did you know that it has been linked to non-Hodgkin lymphoma?"

I didn't. While I was well aware that pesticides had been implicated in this and other cancers, I had not heard that this commonly available consumer product was among them. I googled it immediately, as Tim was talking, and was flabbergasted by what I saw: reports from the International Agency for Research on Cancer linking glyphosate to non-Hodgkin lymphoma, links to articles that debunked such claims, opinion pieces in various outlets about potential litigation being prepared, and so on.

"Interesting," I said. "So how can I help you?"

"We represent patients who have had heavy exposure to Roundup," said Jeff, speaking with a little more urgency. "They've been diagnosed with non-Hodgkin lymphoma, and we believe that Monsanto knew all along that there was a risk."

"We're wondering," continued Tim, "would you be willing to review the evidence, research the association, and let us know what you, as an expert, think?"

I sat back in my office chair and thought for a moment. I appreciated their approach—they weren't pushy and sounded respectful. And I was intrigued. Could there be a massive cover-up involving a product that could be causing cancer in unsuspecting people? Or was I watching too many legal thriller movies?

My curiosity got the best of me. Ignoring how much work I already had on my plate, I said, "Okay. Send me what you've got, and I'll start looking into it. I will also do my own research."

In terms of time management, this might not have been the wisest move. When I accepted the task of reviewing the relationship between Roundup and non-Hodgkin lymphoma, I was an associate professor in the division of hematology and oncology, the director of cancer clinics, and a medical director for the International Program, all at the University of Chicago; I was finishing my second year of business school; my twin boys were completing fourth grade; and my wife was understandably growing tired of my incessant work. Moreover, my parents had settled in the United States after moving here from Syria and were living with us. I was juggling their doctor's appointments and helping them to continue adjusting to a new culture.

Let me rephrase what I said above: taking on this project was way beyond "not the wisest move." It was foolhardy, maybe even insane. But I was interested in finding out whether there was a relationship between Roundup and lymphoma. And if there was one, and if this giant corporation had downplayed it—well, I'd certainly want to support any effort to bring that to light.

That weekend, I began my deep dive into Roundup. I learned that Henri Martin, a Swiss chemist, had been the first person to synthesize glyphosate, the primary ingredient in what would become the most commonly used herbicide in the United States. That was 1950. Twenty years later, John E. Franz, a chemist at Monsanto, independently synthesized glyphosate after his colleagues found that similar chemicals were slightly harmful to plants.

Franz found that glyphosate was a very effective plant killer, which led Monsanto to immediately patent the compound and start selling it in 1974 under the trade name Roundup. Between 1990 and 1996, sales of Roundup increased around 20 percent per year, but what might be called "the Roundup revolution" really took off in 1996, when Monsanto started selling genetically modified soybean seeds that produced crops resistant to the herbicide. Farmers could spray Roundup on their soybean fields, and it would only kill weeds, while the crops continued to grow. "Roundup Ready" soon became Monsanto's trademark for its patented line of glyphosate-resistant crop seeds.

Soon Roundup became the number-one weed killer. As of 2012, its active ingredient, glyphosate, was being used to treat approximately 5 million acres in California for crops like almonds, peaches, cantaloupes, onions, cherries, sweet corn, and citrus. In 2014, more than 18 billion pounds of glyphosate were sprayed all over the world. By 2015, the product was being used in more than 160 countries. Roundup is used most heavily on corn, soy, and cotton crops that have been genetically modified to withstand the chemical. In 2016, genetically engineered crops accounted for 56 percent of glyphosate use around the world. That same year, an analysis by the journal *Environmental Sciences*

Europe found glyphosate use had increased nearly fifteen-fold in just eighteen years. With global use of between 280 and 290 million pounds, glyphosate was the most widely used herbicide in global agriculture, and glyphosate-based products were the second most widely used home and garden herbicides.[1]

Glyphosate is non-selective, which means that it will kill just about any plant it comes in contact with. The plant's leaves absorb the chemical, which passes down to the roots. Glyphosate is usually mixed with surfactants—compounds that are added to help the glyphosate stick to weeds, as opposed to falling immediately into soil—as well as other elements to form the product that Monsanto markets as Roundup. Roundup shuts down the chemical pathway plants use to make proteins they need in order to grow. Without those growth agents, plants die in a matter of days or weeks. Monsanto's patent on glyphosate expired in 2000, making the product available for other companies to sell if they choose to. By 2015, more than 750 products made with glyphosate were available in the United States.[2]

As I did my research, the sheer amount of product being used around the world caught my attention. It sounded like this stuff was being sprayed in gardens and on farm fields from one end of the United States to the other, and internationally. If it was causing any form of cancer, the implications were enormous.

My interest in cancer and oncology research dated back to my 1995 residency at Loyola University in Chicago. I was in the fourth month of my internship when I had to do two oncology rotations back-to-back. The first was in the bone marrow transplant unit, where I took care of very sick patients who were undergoing transplantation, mostly for various blood cancers but some for solid tumors. The second one was on the general oncology ward, and it was there that I developed my passion for oncology as a discipline. I recall walking into the room of a patient early one morning to check on her. She was in her late forties and had advanced-stage ovarian cancer. I was copying her vital signs onto an index card when she opened her eyes and asked weakly, "Do you think I will live until Christmas?"

I was taken aback by this question. I was just starting my residency and had no idea what to say. She saw my discomfort and hesitation and mustered a smile despite the heavy doses of pain medication she was on. "It's okay," she said reassuringly, as if *I* were the one who needed reassuring. "You don't need to answer. But I look forward to seeing your smile tomorrow."

More than twenty-five years later, I can still vividly recall the scene. Her grace and dignity, even as she was battling for her life, affected me deeply. She knew, and she knew that *I* knew, that there was little we could do for her. But as I learned during my rotation in that department, human connections are very important to patients, sometimes more important than another chemo protocol, another surgery, or another pill. While it may sound counterintuitive, it always seemed to me that oncology, despite being a discipline in which many of our patients die, is in some ways the most humane of medical specialties. Perhaps that is because it forces all of us—both patients and those treating them—to come face-to-face with mortality in a way most other areas of medicine do not.

I still feel that way, and I have always felt privileged to treat the patients that I see. More and more of them have survived as our technologies and understanding of the biology of cancer have evolved. Sadly, that wasn't the case with the woman with ovarian cancer who helped me realize my life's calling. That inspiring patient never lived to witness Christmas of 1995.

My interest in non-Hodgkin lymphoma began while I was doing a three-year fellowship at Northwestern University. There, I had the privilege of working with Dr. Steve Rosen, a lymphoma and myeloma expert who also was the Cancer Center director at the time. Steve was and is an amazing scientist, researcher, physician, and human being. To this day, I tell him, every chance I get, that I owe him for the unwavering support that he provided me and for his mentorship over the years. Steve's laboratory work focused on understanding mechanisms of myeloma and lymphoma cell death when the cells are exposed to various compounds and antibodies. I worked in his lab for two years while also seeing patients with various types of non-Hodgkin lymphoma. I think we are mostly shaped by our mentors, and he certainly influenced why I liked seeing and treating patients with lymphoid malignancies.

But many years later, as I was doing my research on Roundup and considering getting involved in the Monsanto case, I knew what Steve would say. Everyone thinks we physicians have the answers, when in fact very often we have none. "I'm not sure what caused your cancer, but let's proceed with treating it" is a line I found myself saying over and over again.

As I went through document after document, though, I began to wonder: Could *this* be an answer for at least some cases of lymphoma? Could glyphosate have been the cause of so much suffering for some of my patients, and so many others?

I decided that maybe this would be the one time I should work with the lawyers and agree to testify in court if needed. At the very least, I decided, I should learn more about this.

A few weeks after our call, and after signing some legal documents, I received several USB drives from the Miller Firm containing confidential documents. Some of these were internal Monsanto communications: emails, memos, things that were never supposed to be read by outsiders. How had these lawyers gotten their hands on them?

My legal education was about to begin. I called Tim, one of the attorneys I had initially spoken with, and asked him how they had managed to acquire these internal documents. He explained that they had been handed over as part of the discovery process in this lawsuit. In discovery, an early stage in lawsuits, lawyers request documents from the opposing party. While these materials were not direct scientific literature, they did reveal how the company was conducting business internally, something that cannot be ignored. Tim shared with me how difficult it had been for them to get these documents and how Monsanto had used every legal argument its lawyers could think of to prevent or delay the release of this critical information.

I was shocked by what *I* discovered from these materials. It turned out that concerns about how this compound could harm humans dated back to the 1980s. In fact, the Environmental Protection Agency (EPA) had stated in 1985 that glyphosate was a possible human carcinogen.[3] But later, the chemical's classification somehow changed, and glyphosate was deemed non-classifiable.[4] Eventually, the compound was classified as non-carcinogenic.[5] It was not clear to me how these shifts had occurred, but certainly it piqued my curiosity and drove me to dive deeper into the trove of information I had been given.

I also learned that in the 1990s, Monsanto had hired a prominent physician named James Parry, who specialized in genotoxicity, or damage that can occur to the DNA of cells, to assess available evidence on whether Roundup was genotoxic. Parry was a respected scientist with a reputation beyond reproach; he was often asked by the UK government to offer his opinion on toxicology issues. Dr. Parry concluded that additional studies were needed to answer this important question, as there was inadequate data on glyphosate's clastogenicity (the ability of a compound to cause mutation and chromosomal breakage). Quite sensibly, in my opinion, he recommended conducting additional studies to determine the potential carcinogenicity of glyphosate.[6] (See figure 1.) Indeed, Parry stated in his report,

Key Issues concerning the potential genotoxicity of glyphosate, glyphosate formulations and surfactants; recommendations for future work.

James M. Parry

Centre for Molecular Genetics and Toxicology
School of Biological Sciences
University of Wales Swansea
Swansea SA2 8PP, UK

Key Questions

1. Is glyphosate an *in vitro* clastogen? Can the positive studies of Lioi *et al* (1998a, 1998b) be reproduced?

2. Is glyphosate an *in vivo* clastogen? Can the positive studies of Bolognesi *et al* (1997) be reproduced?

3. If glyphosate is an *in vitro* and *in vivo* clastogen, what is its mechanism of action and does the mechanism lead to other types of genotoxic activity *in vivo* such as point mutation induction?

4. Does glyphosate produce oxidative damage?

5. Can we explain the reported genotoxic effects of glyphosate on the basis of the induction of oxidative damage?

6. If glyphosate is an *in vivo* genotoxin is its mechanism of action thresholded? Under what conditions of exposure are the antioxidant defences of the cell overwhelmed?

7. Are there differences in the genotoxic activities of glyphosate and glyphosate formulations?

8. Do any of the surfactants contribute to the reported genotoxicity of glyphosate formulations?

Deficiencies in the Data Set

1. No adequate *in vitro* clastogenicity data available for glyphosate formulations.

Figure 1: Portion of the Parry report outlining deficiencies in the dataset that Monsanto had him review.

"If the genotoxicity activity of glyphosate and its formulation is confirmed it would be advisable to determine whether there are exposed individuals and groups within the human population." Parry had outlined that he found no adequate *in vitro* clastogenicity data available for the glyphosate formulations and

2

2.   No bone marrow micronucleus study of glyphosate available using multiple dosing and adequate animal numbers.

3.   No studies available demonstrating the effects of anti-oxidants upon the induction of genotoxic endpoints by glyphosate.

4.   No adequate *in vitro* or *in vivo* clastogenicity data for surfactants used in glyphosate formulations.

Actions Recommended

a)   Provide comprehensive *in vitro* cytogenetic data on glyphosate formulations.

b)   On the assumption that the reported *in vitro* positive clastogenic data for glyphosate is due to oxidative damage determine the influence of antioxidants. Evaluate the clastogenic activity of glyphosate in the presence and absence of a variety of antioxidant activities. Such a study should also incorporate glyphosate formulations to clarify the validity of reports of differences in activity. I recommend that both a) and b) should be undertaken using the *in vitro* micronucleus assay in human lymphocytes. The *in vitro* micronucleus assay would provide a more cost-effective method for evaluating a large number of experimental variables.   *Same as screen chrom ab*

c)   Evaluate the induction of oxidative damage *in vivo* and determine the influence of the antioxidant status of the animals. Determine the exposure concentrations of glyphosate which overwhelm the antioxidant status of tissues.

d)   Perform an *in vivo* bone marrow micronucleus assay with multiple dosing with adequate numbers of animals to determine whether the work of Bolognesi *et al* (1997) can be reproduced.

Figure 1: Continued

similarly, there were no studies on sufactants that were used in these formulations.

But apparently Monsanto didn't see it that way, as the company essentially disregarded Dr. Parry's opinions as a consultant after he reached these conclusions. Moreover, his recommendations—essentially, to do more research—were ignored. "We simply aren't going to do the studies Parry suggests," said one Monsanto

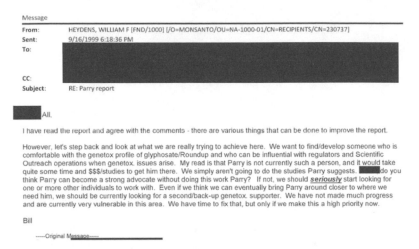

Figure 2: Internal email communication between Monsanto's William Heydens and others within the company discussing in part how Monsanto planned on addressing the Parry report.

scientist. (See figure 2.) It was difficult to believe that Monsanto came to this decision for any reason other than not being happy that Dr. Parry hadn't reached a different conclusion—one that would assert that Roundup had no potential to cause cancer and would strengthen the claims about its safety. Internal email documents showed how Monsanto executives planned to disregard Dr. Parry's suggestions and instead find other scientists who would, essentially, see things differently.

I couldn't believe what I was reading. This wasn't science, and it sounded like a coldhearted refusal to conduct more safety studies. I knew that patients' lives were at stake, and I felt I owed it to them to do my best and seek the truth.

I called Tim Litzenburg. "I'm in," I said.

"That's great, Dr. Nabhan," he replied. "You're going to help us get our clients what they deserve."

2

The First Meeting

About three months after the initial phone call, I found myself driving to meet Tim Litzenburg at a Starbucks not far from my home. Located in a shopping mall with the extravagant name of Plaza del Prado, this Starbucks was near a Tropical Smoothie Café, a Petco, a sushi bar, a McDonald's, and a bagel store. This was one of several coffee shops in my area that I frequented when I needed to concentrate on work in the evening—something hard to do in a house with twin boys.

I arrived at 2 p.m., about thirty minutes ahead of our scheduled appointment, and ordered my usual cup of dark roast. As I was waiting for Tim, I sat back in one of the lounge chairs arrayed around the coffee bar, sipped my drink, and watched the hustle and bustle around me. I saw students poring over textbooks and laptops, people in business suits talking earnestly over the tables as if negotiating some important deal. Nobody seemed to come to this coffee shop to simply drink coffee. It was a place with a purpose.

Every time a man in a suit walked in, I sat up straight, certain that this would be the distinguished attorney from back east. As the minutes ticked away, I wondered if I should order one of the plump buttery scones tempting me from the display case by the checkout counter. No, I resolved. I would wait until Tim arrived and take my cues from him. He probably ate neatly, with a napkin on his lap, befitting what I was sure must be his fancy educational pedigree. Moreover, I felt certain he would conform to my image of a barrister: polished shoes and an impeccably tailored gray suit complete with a pocket square.

Precisely at 2:30 p.m., a scraggly-haired blond guy who looked like he'd just left a Grateful Dead concert entered the Starbucks. He was wearing a T-shirt and jeans—no suit, no briefcase, and no blazer. He was lean and carried a small notebook with a pen attached to it. There would be no scone eating with this guy, I thought disappointedly.

He recognized me immediately, I guess because I was the only one in this Starbucks who *didn't* look like he was cramming for an exam or because he had seen my photo on the University of Chicago website. He walked over and extended his hand.

"Tim Litzenburg," he said. "Nice to see you, Dr. Nabhan."

"Same here," I replied. "But please call me Chadi. I hope you didn't have trouble finding this place. I know it's kind of hidden in this plaza."

"Nothing Uber couldn't resolve," he said with a smile.

Tim ordered an iced drink and we sat down at a quiet corner table.

I had expected that Tim would do most of the talking—that he would tell me more about Monsanto, about the case, about Roundup and the people who alleged they had gotten cancer from it. Instead, the focus was on me.

"Do you live close by?"

"How old are your kids? What grade?"

"I know you've got a lot on your plate. How do you balance your time?"

"Did you get a chance to review all of what we sent you?"

I suddenly had the uneasy feeling that I was already on the witness stand. Was he assessing whether I had the stomach to go through a grueling litigation and trial process?

The most intriguing question he asked was about why I had gone back to business school. Against the objections of my wife, who felt that my other career was quite enough, I had enrolled in Loyola University's Quinlan School of Business two years earlier, in 2014, and was nearing completion of an MBA with a focus on healthcare management. I got the sense that Tim anticipated this question would come up many times in future depositions—and he was right.

As Tim listened and occasionally took notes, I explained that after many years of clinical practice, I wanted to impact patient care from a different angle. While nothing would ever replace the experience of caring directly for patients, I knew there was only so much I could do as a practitioner. It was in administration where I could have an influence on some of the big decisions that would ultimately affect patient care. And to do that effectively and credibly, I thought I would need an MBA. Also, I told Tim, I've always tried to challenge myself, to get out of my comfort zone.

As noble and high-minded as this may sound, the decision hadn't been easy on me or my family. It put a crimp in our finances, because tuition at Loyola University wasn't cheap. I also struggled to balance family life with trying to master accounting and economics.

"I've been reminded that life is about trade-offs," I admitted. Because the program I was enrolled in, an executive MBA with a focus on healthcare management, was geared to people like me who also held a full-time job, most of the classes were on Fridays or Saturdays. But that meant I often wasn't able to attend my kids' activities or family functions. "I rationalized in my own mind that I would make up for lost time once I finished business school," I said with a shrug.

Tim nodded, and then his questions changed direction. "I know you've been doing some research on glyphosate since we first spoke. Now that you've looked at the documents and the literature, what do you think? What's your opinion about the relationship between glyphosate and non-Hodgkin lymphoma?"

I liked his approach—clearly, he wanted my honest assessment, and he wasn't trying to influence my answer.

"I went through everything you and your team sent me," I told him. "And I've also looked at what's been published about glyphosate in the peer-reviewed literature and other outlets. I did my own research and assessment. I wish I could give you a definitive yes-or-no answer. But to be honest, the results are mixed."

Tim didn't bat an eye. "Please go on," he said.

I explained that most of the studies I'd looked at were epidemiological, meaning that they looked at populations, not individuals. Such studies can be enormously valuable but are often inconclusive, for the simple reason that you are looking at disease distribution among large groups of people; it might be very difficult to draw cause-and-effect conclusions with so many variables. That was the case here. The epidemiological evidence of glyphosate's association with non-Hodgkin lymphoma was mixed. I had read some studies that showed a link. Others showed no relation.

"There's a 'but' here, though," I said, and Tim's eyebrows rose. "I found other studies showing that glyphosate *can* cause cellular damage and that it can cause cancer and tumor formation in rats and mice. Also, there were some meta-analyses that showed more convincing associations. With that, the evidence becomes more compelling."

In fact, the International Agency for Research on Cancer (IARC), part of the World Health Organization (WHO), had analyzed all existing peer-reviewed published evidence on glyphosate's potential to cause cancer. In March 2015, members of the IARC met in Lyon, France, to confer and discuss the evidence they had reviewed. The meeting brought together many international, independent scientists who are experts in their respective fields. They concluded that glyphosate is probably a human carcinogen.

This was significant to me, in large part because of the credibility of the IARC, which was founded in 1965 specifically to evaluate compounds that might cause cancer in humans. The potentially hazardous substances that the IARC investigates are those with high levels of human exposure and some preliminary evidence that they might be carcinogenic. Since its inception, IARC has evaluated over 1,000 compounds and determined that about 20 percent are either human carcinogens (group 1) or probable human carcinogens (group 2A). This said to me that the IARC investigators are very careful in their evaluation and in their assessment of the peer-reviewed published literature. Plus, their rigorous process is transparent and well explained on their website, which I also appreciated.

The IARC's findings were later published in *The Lancet Oncology*, one of the world's most prestigious cancer journals.[1] The *Lancet* publishes nothing that has not been rigorously peer-reviewed.

Part of the IARC's transparency involves having outside observers who can witness its evaluation process and see how it reaches its conclusions. I was intrigued to learn that a Monsanto representative had been present during the IARC sessions that evaluated glyphosate. I later found out, through internal Monsanto email communications obtained by the Miller Firm, that the company was preemptively formulating its response to

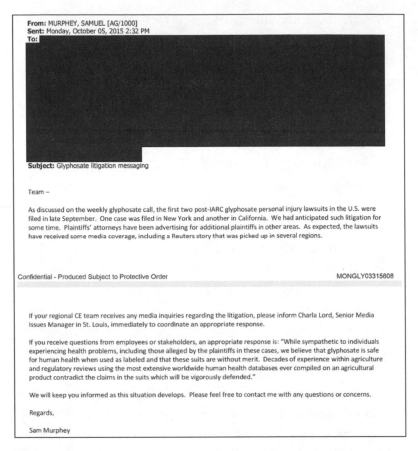

From: MURPHEY, SAMUEL [AG/1000]
Sent: Monday, October 05, 2015 2:32 PM
To:

Subject: Glyphosate litigation messaging

Team –

As discussed on the weekly glyphosate call, the first two post-IARC glyphosate personal injury lawsuits in the U.S. were filed in late September. One case was filed in New York and another in California. We had anticipated such litigation for some time. Plaintiffs' attorneys have been advertising for additional plaintiffs in other areas. As expected, the lawsuits have received some media coverage, including a Reuters story that was picked up in several regions.

Confidential - Produced Subject to Protective Order   MONGLY03315608

If your regional CE team receives any media inquiries regarding the litigation, please inform Charla Lord, Senior Media Issues Manager in St. Louis, immediately to coordinate an appropriate response.

If you receive questions from employees or stakeholders, an appropriate response is: "While sympathetic to individuals experiencing health problems, including those alleged by the plaintiffs in these cases, we believe that glyphosate is safe for human health when used as labeled and that these suits are without merit. Decades of experience within agriculture and regulatory reviews using the most extensive worldwide human health databases ever compiled on an agricultural product contradict the claims in the suits which will be vigorously defended."

We will keep you informed as this situation develops. Please feel free to contact me with any questions or concerns.

Regards,

Sam Murphey

Figure 3: Internal Monsanto email communication showing their anticipation of litigation trials.

the forthcoming IARC report. When it was published in March 2015, Monsanto was prepared to counter.

Indeed, after the IARC concluded that glyphosate was probably carcinogenic, Monsanto mounted a campaign to counter the IARC's findings. I wouldn't have expected them not to challenge the results. But other documents made available through court proceedings showed that this went beyond a reasonable discourse: the corporation deployed an entire team of scientists and regulatory experts to orchestrate a response *discrediting* the

IARC's conclusions. Monsanto clearly had anticipated that the IARC's findings wouldn't be in their favor and that there would likely be a backlash after the report was released, including possible litigation (see figure 3). Indeed, as we will see, subsequent internal memos confirmed that the corporation recognized substantial vulnerabilities in their own data.

What shocked me most was the extent to which Monsanto failed to acknowledge the possible danger with its product and its unwillingness to warn the public. Even if the corporation was not convinced of the validity of the IARC's evaluation, they should at least have acknowledged the report and then offered their rebuttal. Instead, they launched a vigorous campaign to show how safe glyphosate is. That decision certainly backfired, considering that Monsanto was now facing one of the largest product liability lawsuits in history.

I also recognized a historical parallel here. The way Monsanto denied the evidence reminded me of how tobacco manufacturers had spent decades and billions of dollars denying the fact that smoking could cause cancer. We all know how that story ended: now every consumer can read the warnings from the surgeon general about the cancer risks from smoking.

Although I would certainly keep an open mind when contradictory evidence emerged, to me it now seemed clear that glyphosate could cause non-Hodgkin lymphoma, at least in some individuals. Also, let's be clear: the evidence wasn't showing (nor have I ever claimed, under oath or not) that *everyone* who uses Roundup will get cancer. But instead of acknowledging that some people might, and maybe engaging in a legitimate debate on how to prevent more people from becoming ill, Monsanto chose instead to challenge and attempt to discredit the scientists who disagreed with them.

Three hours after we'd started talking, and long after my cup of coffee had gone cold, my meeting with Tim concluded. "Glad you're on board, Chadi," he said as he closed his notebook. "You're going to help a lot of unfortunate people."

Though I was still unaware of the scope of this lawsuit or how big these cases would become, Tim and I were on the same page, and I felt more energized and willing to contribute, because this struck at the heart of something I've frequently encountered in my many years as a physician: the "why" question, "Why did I get cancer?"

I'd heard it from almost every patient diagnosed with non-Hodgkin lymphoma. Usually I had no answer, as most lymphomas are of unknown cause. But in the few scenarios where we were able to identify a cause, I saw how helpful that knowledge was for patients and families. At least they had some answers.

Furthermore, if we could identify risk factors that were modifiable or avoidable, then we could make an enormous impact on patient care. Again, I make the comparison to tobacco consumption. Even when a smoker develops cancer, counseling them about how to stop smoking can be valuable. It can prevent the development of another cancer and reduce the chances of the current cancer getting worse. Could something similar, such as a warning label on Roundup, be one of the outcomes of this litigation? I certainly hoped so.

Over the next few days, Tim sent me the transcribed depositions that were available at the time. A deposition is out-of-court testimony provided under oath, and it's often videotaped. I spent many nights reading the transcriptions, and it quickly became apparent to me that Monsanto had likely engaged in a practice known as "ghostwriting," meaning that they influenced the writing of scientific articles (even drafting large portions of these articles) without acknowledging that they had done so. To make these articles appear independently researched and written, Monsanto engaged researchers to co-author these pieces and publish them in peer-reviewed journals.

While I had questions and, already, some strong opinions about the legal and ethical aspects of this case, my job was not to plan a legal strategy, so I stuck to my task: investigating how patients can develop non-Hodgkin lymphoma from this agent and whether the lymphoma diagnosed in the patients I was reviewing was attributable to glyphosate and Roundup. The implications here were enormous, considering how ubiquitous the use of glyphosate is across the United States and the entire world. What if glyphosate caused other illnesses beyond non-Hodgkin lymphoma? How much exposure was needed to put an individual at risk? Was glyphosate present in our food, and were unsuspecting people being exposed to it?

I had been asked to provide a written report of my findings. (To the Miller Firm's credit, they never offered to help me write the report.) Pursuant to the Federal Rules of Civil Procedure that govern discovery, my report would be shared with the Monsanto side. I figured they wanted to see what I wrote so that they could prepare their strategy and questions for my first deposition.

I also knew that as soon as they read what I had to say, I was going to be seen as a hostile witness and as a threat to Monsanto. A $66 billion corporation was going to have me in their sights.

Soon, I knew, I might have to convince a judge and a jury. But I first had to convince Lama, my wife, that I needed to take this on—something it took me months to address. I'd started doing the preliminary research without telling her what I was doing. Between finishing up my MBA and starting to look into the Monsanto case, I was staying up late working most nights, and my temper was shorter than usual. One evening I'd been trying to read a particularly complex epidemiological study on glyphosate and barked at her because I felt she had the TV volume up too loud. I apologized right away. "Sorry to have such a short fuse," I told her. "I'm working on something big that has the potential to help lots and lots of people." She shook her head and left the room to watch her show on her mobile device.

Eventually, though, I had to come clean. So, one evening after dinner I said the words that every spouse dreads hearing: "I need to talk to you." Slowly, and hemming and hawing a bit, I told her what had transpired over the preceding few months and that I'd committed to working on this case.

"I really think I can help some people here," I concluded. "This is a huge case."

She looked down at her lap for a while before speaking. "Chadi, I think it's wonderful that you want to help people," she finally said. "But could you please not do this? Could you please reconsider?"

Lama offered compelling reasons: "You don't have the time. Your kids need you. We all need you." I had no answer to that, and she continued. "You're finally finishing business school. We supported your decision to do that, and we've waited two years for that day so we can have you back. And now you're off to something else." She was getting upset, and her voice got louder. "How is this fair to anyone?" She paused, and when she spoke again, I heard fear in her voice. "And what if something happens to you? What if these people go to the extreme to hurt you if you and the lawyers are going after them?"

I gave her a hug. "You don't have to worry about that," I said. Trying to ease the tension a little, I joked, "I think you've been watching too many episodes of *Law and Order*." I reassured her that I was focusing on the science and that Monsanto was not going to go after me personally. I told her that I was just one of many expert witnesses taking part in this case. "This is just a civil lawsuit," I concluded. "These types of things take place all the time."

I'm not sure if I was trying to convince her or myself, but despite all the reasons I'd given her why there was no cause for concern, I certainly became more careful when walking alone at night. Eventually I concluded that maybe I, too, had watched a few too many drama series on TV. But it didn't stop me from looking over my shoulder in parking garages or on dark, empty streets for many months to come.

3

The EPA

I was convinced that glyphosate could be implicated as probably carcinogenic. Now it was time for me to come face-to-face with those who were going to be working actively to discredit me and the other experts.

It was August 2017 when I met with some of the Monsanto lawyers for the first time. My legal vocabulary was rapidly expanding, and now I learned another new term: this was my deposition on "general causation." For purposes of product liability, Tim explained to me, general causation refers to the capability of a substance to cause the specific harm in question, and in this litigation it's cancer. Specific causation asks the question of whether a potentially carcinogenic material caused cancer in a particular individual.

In other words, in this first deposition I'd be questioned about whether glyphosate and Roundup could cause cancer in general, not whether it caused cancer in the case of any specific individuals (such as the plaintiffs the Miller Firm or others would be representing in the upcoming trials).

I was eager to learn about these fine points of law. At one point I had toyed with the idea of going to law school, but Lama immediately shot that one down. "Chadi, you've got to be kidding me," she said. "Law school? For a doctor and an administrator who is raising a family?" She had an excellent point. Nonetheless, I continued to watch legal shows and movies when I had any downtime. Something about how lawyers strategize their cases intrigued me. But watching law dramas had been the extent of my law school curriculum—until now.

The August deposition would give the Monsanto legal team the opportunity to explore the basis of my opinion that Roundup and its active ingredient, glyphosate, could cause non-Hodgkin lymphoma, well before we got into a courtroom. We would mostly be talking about the research that had been done on the relationship between glyphosate and lymphoma. I was certain that one of the core arguments Monsanto lawyers would make was that the EPA had sided with them when they concluded that glyphosate was not carcinogenic, and that this was by itself sufficient to exclude any potential liability if users developed cancer. If the EPA had blessed Roundup, then what was all the fuss about?

To really understand the underpinnings of this case and the EPA's role in it, I believe we have to go back to before there even was a federal agency mandated to protect the environment. In 1962, marine biologist and nature writer Rachel Carson published a book, *Silent Spring*, in which she documented the adverse effects pesticides were having on people and the environment. The book was a revelation to many and is now widely considered to have helped spark the modern environmental movement. In *Silent Spring*, Carson made a compelling case that the chemical industry often distorted science and spread misinformation in order to avoid any harm to the image or sales of their products. She exposed the marketing ploys for these pesticides that led to their wider use but avoided the subject of any unintended consequences of that use.

The book was an immediate bestseller and quickly gained national media attention. Despite fierce efforts by the chemical industry to discredit Carson and her book (she was called a crank and a Communist, among other things), *Silent Spring* went on to sell more than 6 million copies in English and has been translated into thirty different languages.

After the book came out, public opinion began to turn. Ten years after its publication, the United States banned the use of dichloro-diphenyl-trichloroethane (DDT) for agricultural purposes. The ban was announced by William D. Ruckelshaus, the administrator of a new agency that had been created by President Richard Nixon to respond to the concerns—and a new and growing environmental movement—sparked in part by *Silent Spring.*

The agency had begun in the summer of 1969 as something called the Environmental Quality Council. That spring, the newly inaugurated president had walked the beach in Santa Barbara, California (wearing a suit and black oxfords, or so the story goes), after an oil spill had spoiled that idyllic setting. Nixon promised action. "What is involved," he said to reporters as protestors chanted nearby, "is the use of our resources of the sea and the land in a more effective way, and with more concern for preserving the beauty and the natural resources that are so important to any kind of society that we want for the future. I don't think we have paid enough attention to this. . . . We are going to do a better job than we have done in the past."[1]

The Environmental Quality Council would be one of his first environment-related actions. In a 2017 story on the history of the agency's development, *Time* magazine described the new organization as "a Cabinet-level advisory group designed to co-ordinate governmental action against environmental decay at all levels, create new proposals to control pollution, and foresee problems."[2]

Shortly afterward, Congress passed the Environmental Policy Act of 1969, one element of which was the creation of a Council on Environmental Policy, "empowered to review all federal activities that affect the quality of life and make reports directly to the President," per the magazine. A year later, President Nixon announced the sweeping powers of the new agency he wanted to create. The mission of his new Environmental Protection Agency would focus on:

- Establishing and enforcing environmental protection standards consistent with national environmental goals.
- Conducting research on the adverse effects of pollution and on methods and equipment for controlling it; gathering information on pollution; and using this information to strengthen environmental protection programs and recommend policy changes.
- Assisting others, through grants, technical assistance, and other means, in arresting pollution of the environment.
- Assisting the Council on Environmental Quality in developing and recommending to the president new policies for the protection of the environment.[3]

Today, the EPA employs more than 13,000 people, including engineers, scientists, information technologists, and other specialists, charged with conducting research and assessments of the environment and to enforce environmental laws.[4] But while the EPA has had many notable achievements, including the banning of DDT, the fight against acid rain, and limiting auto emissions, it seemed rather bullish on glyphosate—something I discovered after the countless hours I spent reviewing the science and internal Monsanto documents.

In 1974, just four years after the EPA was established, glyphosate came onto the market. That year alone, 1.4 million pounds of Roundup were sprayed on farm and ranch lands across the United

States. Glyphosate use began to mushroom in the 1990s when the US Department of Agriculture approved Monsanto's request to market corn, soy, and cotton seeds that had been genetically engineered to resist Roundup. By 2014, Americans were using 276 million pounds of the stuff. In the United States, the EPA has registered glyphosate for use on more than a hundred crops, including wheat, rice, oats, barley, and alfalfa. In California alone, more than 11 million pounds of glyphosate were used on crops in 2015, including almonds, avocados, cantaloupes, oranges, grapes, and pistachios. In 2015 (the year the IARC report was released), Monsanto's sales of agricultural productivity reportedly brought in $4.76 billion, much of that likely from the sale of glyphosate used on fields planted with the company's glyphosate-resistant GMO seeds like Roundup Ready soybeans.[5]

Back in 1975, when glyphosate entered the market, the young EPA did not have the human or scientific resources to conduct the necessary investigations to determine the safety of every single pesticide in use. It took them over a decade to finalize their analysis of glyphosate. During that period, however, multiple red flags had been raised by various EPA employees, and many of them seemed to have been ignored. In an August 1978 memo, EPA scientist Krystyna Locke raised concerns about a Monsanto study of glyphosate in which scientists from the contract lab Monsanto had hired failed to record what happened in the experiment. Locke quoted Monsanto's scientist Robert Roudabush, who defended the study this way: "The scientific integrity of a study should not be doubted because of the inability to observe all primary recording of data."[6] In other words, Roudabush and Monsanto seemed to be saying that the EPA should not be concerned by the absence of data. Their message, in essence, was "Would we lie to you?"

It turned out that tests done on glyphosate to meet its EPA registration requirements had been carried out by a lab associated

FROM
(NAME—LOCATION—PHONE)   Dept. of Medicine & Environmental Health   G.J. Levinskas, G2WF 4-8809

DATE   :   April 3, 1985   cc.   G. Roush, Jr., M.D.

SUBJECT   :

REFERENCE   :

TO   :
                   T.F. Evans

                   The following item of information is in addition to those
                   included in the current monthly report.

                   Senior management at EPA is reviewing a proposal to classify
                   glyphosate as a class C "possible human carcinogen" because
                   of kidney adenomas in male mice. Dr. Marvin Kuschner will
                   review kidney sections and present his evaluation of them to
                   EPA in an effort to persuade the agency that the observed
                   tumors are not related to glyphosate.

                                              George J. Levinskas

                   GJL/sfd

Figure 5: Internal Monsanto memo stating that Dr. Marvin Kuschner will "persuade the agency that the observed tumors are not related to glyphosate." This was written before Kuschner ever reviewed any pathology.

the highest dose) developed kidney adenomas, a type of tumor that typically is benign but can become cancerous. (See figure 4.) None of the unexposed mice grew any tumors. Based on these findings, scientists at the EPA declared glyphosate to be carcinogenic. Not unexpectedly, Monsanto fired back, claiming that these tumors had occurred by chance and were not related to glyphosate.

The company decided to retain a prominent pathologist, Dr. Marvin Kuschner, who at the time was the dean of the medical school at the State University of New York at Stony Brook (now Stony Brook University), to review the pathology slides. An internal Monsanto memo that was among the papers the Miller Firm obtained implied that the results were preordained: the pathologist would find no tumors related to glyphosate. (See figure 5.)

Kuschner's report indicated that he found a kidney tumor in one of the mice that were not exposed to glyphosate. It's worth noting that no other pathologist was able to find the tumor that he reported. Nevertheless, Kuschner's assessment was significant, as it eliminated glyphosate as a culprit in the other tumors.

So, what really happened here? I'm sure a lot of reporters and researchers at the time would have liked to know more about Kuschner's findings, and why his results differed from those of the previous studies. No dice. Monsanto reported the Kuschner findings to the EPA in the fall of 1985 as a "trade secret," so the public was not allowed to access the report.

As Monsanto had hoped, this finding cast doubt on the original EPA classification of glyphosate as carcinogenic, and in 1987 the agency subsequently changed its classification of glyphosate to class D, meaning that it was no longer deemed a carcinogen.[11] The EPA, however, did ask Monsanto to repeat the mouse study so that a definitive conclusion could be reached. Monsanto never followed through, however, and after a rather unproductive dialogue, in which the agency asked for a repeat study to properly assess carcinogenicity and the company refused for a multitude of reasons, the agency finally blinked. In the summer of 1989, the EPA dropped its request for a repeat study. Why? I was unable to find out.[12]

In June 1991, the EPA reclassified glyphosate as group E, meaning there was no evidence of carcinogenicity in humans.[13] This was a victory for Monsanto. The prior classification, group D, was for agents without adequate data either to support or refute human carcinogenicity. Class E was more decisive. This was the EPA's version of the Good Housekeeping seal. Glyphosate was getting a big thumbs-up.

What appeared to me to be a cozy relationship between the EPA and Monsanto continued over the years, even after the

IARC report was released. In February 2015, a month before the IARC report came out, the Agency for Toxic Substances and Disease Registry (ATSDR), a federal public health agency that is part of the Department of Health and Human Services (HHS),

Message
_____

From:	HEYDENS, WILLIAM F [AG/1000] [/o=Monsanto/ou=NA-1000-01/cn=Recipients/cn=230737]
on behalf of	HEYDENS, WILLIAM F [AG/1000]
Sent:	4/28/2015 2:52:30 PM
To:	JENKINS, DANIEL J [AG/1920] [/o=Monsanto/ou=NA-1000-01/cn=Recipients/cn=813004]
CC:	LISTELLO, JENNIFER J [AG/1000] [/o=Monsanto/ou=NA-1000-01/cn=Recipients/cn=533682]
Subject:	RE: Glyphosate IARC Question

Dan,

Wow! - that's very encouraging. Thanks for the news update.

Regarding the sarcomas Jess mentions in Cheminova's mouse study, I'm assuming he is talking about the Haemangiosarcomas in high dose males (1000 mg/kg/day, the limit dose) and low numbers (1-3) of histiocytic sarcomas 'spattered' across all dose groups. These were discussed in the 2004 WHO/FAO JMPR documents which states: "Owing to the lack of a dose-response relationship, the lack of statistical significance and the fact that the incidences recorded in this study fell within the historical ranges for control, these changes are not considered to be caused by administration of glyphosate."

Bill

From: JENKINS, DANIEL J [AG/1920]
Sent: Tuesday, April 28, 2015 9:33 AM
To: HEYDENS, WILLIAM F [AG/1000]
Cc: LISTELLO, JENNIFER J [AG/1000]
Subject: RE: Glyphosate IARC Question

Hey- cc'ing Jen

So...Jess called me out of the blue this morning:

Figure 6: This internal Monsanto communication appears to show a relationship between Monsanto and EPA deputy director Jess Rowland. There was communication between Monsanto and Jess Rowland while he was at the EPA. (*continues*)

"We have enough to sustain our conclusions. Don't need gene tox or epi. The only thing is the cheminova study with the sarcoma in mice- we have that study now and its conclusions are irrelevant (bc at limit dose...?). I am the chair of the CARC and my folks are running this process for glyphosate in reg review. I have called a CARC meeting in June..."

Also, Jess called to ask for a contact name at ATSDR. I passed on Jesslyn's email. He told me no coordination is going on and he wanted to establish some saying "If I can kill this I should get a medal". However, don't get your hopes up, I doubt EPA and Jess can kill this; but it's good to know they are going to actually make the effort now to coordinate due to our pressing and their shared concern that ATSDR is consistent in its conclusions w EPA.

Dan Jenkins
U.S. Agency Lead

Regulatory Affairs
Monsanto Company
1300 I St., NW
Suite 450 East
Washington, DC 20005

Office: 202-383-2851

Cell: 571-732-6575

From: HEYDENS, WILLIAM F [AG/1000]
Sent: Monday, April 27, 2015 1:20 PM
To: JENKINS, DANIEL J [AG/1920]
Subject: RE: Glyphosate IARC Question

That would be great, Dan.

Figure 6: Continued

announced that it was planning to issue a toxicology profile of glyphosate later that year. It's noteworthy that the ATSDR assessment wasn't published until early 2019—three and a half years later.[14]

UNITED STATES DISTRICT COURT

NORTHERN DISTRICT OF CALIFORNIA

IN RE: ROUNDUP PRODUCTS LIABILITY LITIGATION	MDL No. 2741 Case No. 16-md-02741-VC
This document relates to: ALL ACTIONS	**PRETRIAL ORDER NO. 19:** **ROWLAND DEPOSITION TOPICS** Re: Dkt. No. 259

This order reflects the Court's verbal ruling in the off-the-record conference call held on

April 24, 2017. Mr. Rowland is ordered to answer questions about the identities of the

companies for which he has done consulting work since leaving the EPA and questions eliciting

a very general description of the projects he has worked on.

IT IS SO ORDERED.

Dated: April 24, 2017

VINCE CHHABRIA
United States District Judge

Figure 7: Judge Chhabria mandates that Jess Rowland answer questions from plaintiff counsel.

While it might not be fully clear why there was a significant delay in publishing the ATSDR's assessment of glyphosate, some have proposed that Monsanto had a friend in Washington, DC, who was ready to help. Records that have been made public suggest that Jess Rowland, a deputy director in the EPA's Office of Pesticide Programs at the time, may have been in regular communication with Monsanto officials and had told Monsanto executive Dan Jenkins in reference to the HHS investigation, "If I can kill this, I should get a medal."[15] (See figure 6.)

Plaintiff counsel, including the Miller Firm, made a significant effort to depose Jess Rowland, while Monsanto worked to block that possibility. Ultimately, Rowland was deposed, and Judge Chhabria ordered Rowland to testify about his time at the

EPA as well as give limited testimony about his consulting work after leaving the EPA. In fact, during his deposition, Rowland's attorney was instructing Rowland not to identify the chemical companies he was consulting for after he left the EPA. Mike Miller, who was deposing Rowland, called Judge Chhabria during the deposition; Judge Chhabria ordered Rowland to identify those companies and to answer Miller's questions on the matter. (See figure 7.) When addressing the Rowland deposition, Judge Chhabria stated, among other things, "The Court will use this guidepost to consider the plaintiffs' efforts to take Jess Rowland's deposition and to compel production of documents relating to his work. Accordingly, the Court is of the tentative view that the testimony and documents the plaintiffs seek from Rowland would be appropriate, but that further discovery from EPA officials would not be. However, the Court will consider any further arguments from the EPA in a motion to quash (to be filed no later than March 28, 2017) before making a final decision. In the event of a motion to quash, the Court will also consider any arguments regarding the application or validity of the EPA's Touhy regulation."[16]

Learning about these emails and communications made me more intrigued and puzzled. But, let me stick to the science and ignore the noise. To me, the association between glyphosate and non-Hodgkin lymphoma was clear, and I had no problem saying that to a judge, a jury, or Monsanto's lawyers.

I was about to get an opportunity to do just that, as the deposition with the Monsanto lawyers would be taking place in the building where I had my office. At the time, I was employed by Cardinal Health and worked out of their facility in Waukegan, Illinois. The deposition took place in a large conference room with an oval-shaped dark-brown table positioned in the middle. The videographer set up the equipment and checked to see that my image was centered in the frame. A court reporter was there to swear me in. Two Monsanto lawyers were seated to my left. They were well-dressed, calm, and collected. I tried to study their body language, just as, I am certain, they were studying mine. Two plaintiffs' lawyers, one of whom was Tim Litzenburg, were on my right.

A small microphone was clipped on my dark-blue tie, and we started the deposition at 10 a.m. on August 23, 2017. While I was confident in the conclusions I had reached after reviewing the literature on glyphosate, I'll admit that I was anxious. I had no idea what Monsanto's lawyers might ask or say.

By agreement between the plaintiff and defense attorneys, it had been determined that seven and a half hours would be allocated for the defense lawyers to ask me questions. That was far longer than any scientific presentation I'd ever delivered, and far longer than any meeting I'd ever sat through, but apparently that was a standard length for such depositions. Still, I felt like a middle-distance runner about to compete in his first marathon.

These depositions are carefully timed, so the attorneys on my side could immediately call the whole proceeding to a halt once that magic seven-and-a-half-hour mark was reached.

My goal was to share what I had learned and explain the evidence of glyphosate's impact on patients. I had agreed to get involved with this case because of its potential impact on patients and society, and I was determined to speak to the science. I also knew that I needed to keep my cool if they tried to attack or discredit me, which Tim had warned me they might do.

One of the Monsanto lawyers, Kirby Griffis, started out just as we had thought he might. He wanted to discuss my MBA, artfully trying to position me as a "suit," a manager or executive who worked in the C-suite of a hospital or research facility but had nothing to do with hands-on science. What jury would believe anything a guy in management might have to say about the veracity of scientific research?

"You got an MBA in support of your current role as a business-person; is that right?" asked Griffis cordially.

What did he mean by "businessperson?" I wondered. *I am a "doctor person."*

Chadi Nabhan: *I actually decided to go back to get my MBA when I was at the University of Chicago as the director of the Clinical Cancer Center and cancer clinics. I wanted to better understand the economics, business, accounting, which will help in my role at the time. So, I got my MBA focusing on healthcare management. My goal was to help more patients at a larger scale. And, you know, this opportunity came along after the fact that I was already on the MBA. I went back to school in August 2014. I was still at the University of Chicago. My goal was just to better understand business of medicine. I think with what's going on in medicine, it is very important for physicians to take lead into understanding business and the impact on patients.*

Kirby Griffis: *It reflected a shift in your interest from patient care to a broader administration and business side and serving medicine though that means. Is that fair to say?*

CN: *No, I don't think it's fair to say. I think delivering patient care is both sides; right? I mean, I think when you take care of patients in clinic, you still have to bill for services. You have to run a business. So, being able to deliver quality care to patients implies that you know how to run your business.*

KG: *Yes, sir. And you're focused now on the business side?*

CN: *I am focused on the business side, but I don't think it's irrelevant to patient care.*

The focus of the questions soon shifted to how I had reached my general conclusions about the association between glyphosate and non-Hodgkin lymphoma. Griffis attempted to downplay the epidemiological evidence, highlighting only the negative studies and critiquing the positive ones. I was questioned about the content of my report. I did my best to explain that we needed to take any and all epidemiological studies seriously since lives were at stake. I spoke about the significance of the IARC findings. Monsanto's lawyer seized on that, attempting to show that I relied too heavily on the IARC report.

Griffis was polite, considerate, and calm. But I had to listen carefully to the way he phrased each question, because I knew he was trying to elicit a response that would make me look bad, or at least like someone unqualified to assess this material.

KG: You, sir, are not an epidemiologist, and you never were one; is that right?

CN: Correct.

KG: You're not a toxicologist, and you never were one; right?

CN: Correct.

KG: You don't call yourself an expert in the mechanisms of carcinogenesis; is that right?

CN: I am not an expert in the mechanisms of carcinogenesis. I can understand the papers that discuss carcinogenesis, and I try my best to look into how this might imply clinical decisions in clinical care.

When I reviewed the transcript later, I realized that he had gotten the better of me on that exchange.

Based in Washington, DC, and a graduate of the University of Virginia Law School, Griffis was going to attempt to take apart the IARC report that was part of the mountain of evidence that had convinced me. In this exchange, he was trying to establish the legitimacy of some of the data that Monsanto had provided to the EPA, but which was not published in the peer-reviewed literature. He showed me some of this data, and then asked:

KG: *You know that the IARC did not review this data in any form; correct?*

CN: *I don't know if the IARC reviewed this particular data. What I know is that the IARC concluded that there's sufficient evidence based on animal studies that there is carcinogenicity.*

KG: *You know that IARC has a policy of not reviewing anything unpublished, correct?*

CN: *I think it's fair to review only published data.*

KG: *And you know that none of this data was published except in the form of this article, correct?*

CN: *Whether they reviewed this particular paper or not, I don't know, but I know that their review collectively demonstrated that the animal studies that they looked at had sufficient evidence to establish carcinogenicity.*

KG: *And you know that it's IARC's policy not to review unpublished studies regardless of their quality, correct?*

CN: *Studies of good quality should be published. So, if they're not published, then why should they be reviewed?*

I felt strongly about that, and it probably came across in the tone of my voice. Peer review is the gold standard in research and scientific publication. Why should that standard be lowered to suit Monsanto's interests?

At that point Griffis changed lanes and began asking me about "registration studies," which are defined by *Legal Insider* as clinical trials that are intended to obtain sufficient data to support the filing of an application for regulatory approval.[17] He was also defending the fact that some of these studies were favorable to Monsanto's view but were not published in the peer-reviewed literature because they were "intellectual property." Monsanto wanted to be able to argue that these unpublished studies should be given weight, too.

KG: *Do you know that registration studies are considered their intellectual property and that, if they were published in their entirety, then another registrant could just submit them to the EPA and get a generic form of glyphosate registered thereby?*

Tim Litzenburg: *Object to form.*

CN: *Almost all registration studies for cancer therapies have to be published in peer-reviewed journals. So, I'm not sure if there's a different thing for compounds like this, but pretty much every drug that has been approved for the treatment of cancer through registration trial has been published in a peer-reviewed journal.*

KG: *Okay. And that's not the case for herbicides. Did you know that?*

CN: *I did not but, in my opinion, if there is literature that is sufficient and compelling, then it should be subject to a peer-review process and the rigor of peer review and get published. There is no reason not to get published.*

The IARC's policy of not using unpublished studies was largely irrelevant to the real issue. The fact was that numerous published studies had found a link between glyphosate and lymphoma. That was the central question of this trial, and I was not going to be steered away from it.

The deposition ended a little after 5 p.m., and I was exhausted. Still, I felt that I had stood up to their questioning and stuck to my guns. Tim and the other plaintiff attorney appeared pleased, but I couldn't help wondering if they were putting on a happy face to encourage me, since this was my first litigation experience.

Although I'd passed this first, crucial test, another key moment loomed. I'd read the documents showing that glyphosate could cause lymphoma, and that Monsanto did not acknowledge that fact. Soon, though, I was going to look the reality of this right in the eye: I was about to meet the first plaintiff himself.

4

Meeting Mr. Johnson

In the weeks that followed, I reviewed the medical records of Dewayne Anthony "Lee" Johnson, whose case was slated to be the first actual Roundup lawsuit litigated in front of a jury. Lee, as he was known, was a man from California who worked as a groundskeeper for a school district. Two years after he started working there, and only in his early forties, he developed a rash that was eventually diagnosed as cutaneous T-cell lymphoma, a rare form of non-Hodgkin lymphoma that affects the skin and can potentially spread to internal organs.

Even before meeting him, and from reading his medical records, I could tell the extent of Lee Johnson's suffering and what he was going through. His lymphoma had disfigured his skin and was causing significant emotional distress. He was not responding well to the therapies his doctors had prescribed. The prognosis was not good. Because of that, Lee's case was a "preference" trial, meaning that it was expedited; he would have his day in court before his health status worsened further.

Given what this man was going through, I was really galled to learn that Lee said he had called Monsanto twice asking whether his lymphoma could possibly be related to his use of Roundup. He never received a call back. After I had a chance to review the records, we arranged for him to come meet me in person in my Chicago office so that I could get a better history and have a discussion with him about his exposure and his current treatment.

To better understand Lee's particular case, we need to look at this disease in general. Non-Hodgkin lymphoma is a form of cancer—an uncontrolled growth of cells in our bodies. Normal cells reproduce and grow at a certain predictable rate, and their life span depends on their origin and other factors. With cancer, this process is disrupted, and the cell cycle is thrown out of whack. Environmental factors are one of the things that can cause this unregulated cell growth.

Cancer is often labeled based on the site of origin. So, if the cancerous cells originate from the breast, then the patient has breast cancer. If they evolve in the colon, then the patient is labeled as having colon cancer, and so forth. There are occasions when we don't know where the cancer cells have originated from, and patients are then labeled as having cancer of "unknown primary."

Non-Hodgkin lymphoma is a form of cancer that originates in lymphocytes, which are a subtype of white blood cells. The disease bears the name of British physician Thomas Hodgkin, who in 1832 at Guy's Hospital in London first described abnormalities in the lymphatic system—what we now call Hodgkin lymphoma. Non-Hodgkin lymphoma is a different form of cancer that also affects the lymphatic system.

Lymphocytes are produced from the bone marrow, circulate in the blood, and reach the lymph nodes (glands) in our neck, armpits, groin, and other areas. These lymphocytes usually defend the body from foreign pathogens and are also important in the way our bodies fight cancer. There are two types of lymphocytes: B-cells and T-cells. Simply put, they work together as a sort of a team to fight foreign pathogens.

But when the lymphocytes become cancerous, they grow uncontrollably in the lymph nodes in these areas. Depending

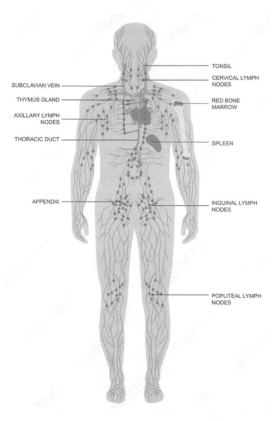

SUBCLAVIAN VEIN

THYMUS GLAND

AXILLARY LYMPH
NODES

THORACIC DUCT

APPENDIX

TONSIL

CERVICAL LYMPH
NODES

RED BONE
MARROW

SPLEEN

INGUINAL LYMPH
NODES

POPLITEAL LYMPH
NODES

Figure 8: Simplified diagram of the human lymphatic system.

on where these lymphocytes grow, the type of lymphoma will be different (see figure 8). There are almost sixty types of non-Hodgkin lymphoma. Each of these can have its own prognosis and treatment.

The majority of lymphomas originate in the B-cells. There are about forty subtypes of B-cell lymphomas, and their classification continues to evolve, based on growing knowledge of the disease and our ability to determine the prognosis of each subtype.[1]

Occupational exposures have also been proposed as possibly causing lymphomas, such as in farmers exposed to heavy pesticide use. Could the commonly used Roundup have caused lymphomas in some of those who used it—say, someone who was a groundskeeper, like Lee Johnson? I believed it could.

But that's not an easy case to make, in medicine or in law, and it goes back to the variation in cancers and how they develop. Often, we don't have a logical explanation. "My father never smoked, he took care of himself, and he didn't work in a job where he was exposed to chemicals. So why did he get cancer?" That's a typical question I'll hear in my line of work. I wish I had an answer for the distraught family members who ask me, but I don't. However, it is a reminder of why it's important to investigate every case individually.

Despite its often-puzzling origins, we do know that non-Hodgkin lymphoma is generally a disease that afflicts older adults. The median age at diagnosis is about sixty-five. Still, I have seen many younger patients afflicted with this disease, and typically no identifiable cause is found. This, however, never prevented me from asking critical questions in order to make progress toward determining some of the causative factors. Even if knowing the cause doesn't affect how we treat a given case of lymphoma, that knowledge could provide a certain sense of closure for some patients and their families. I hoped to be able to do that for Lee Johnson.

My meeting with Lee was planned for the late morning on a mid-October day in 2017, a couple of months after my general causation deposition. I had been worried about whether he would be able to come, as my initial review of his records suggested that he might not survive for long. However, I learned that he had flown from San Francisco to Chicago the night before we met, rented a car, and was staying at a nearby hotel. This was a pleasant surprise, as it implied he was doing better than I would have expected from reading his records.

As I waited for him that morning, I began to become concerned that he would have a hard time finding my office. At the time, it was in a business park that had been designed with all the charm and originality of a cement block, which, in fact, was what pretty much all the buildings in the park were made of. Tim had given me Lee's number, so I called him. "Just want to make sure you're not lost. I'll wait for you outside the building at the entrance. It's tricky to find."

"No problem," he said. Lee's deep, smooth baritone sounded like it could have belonged to a late-night radio deejay.

Fifteen minutes later, I spotted him walking across the parking lot toward the entrance. When he arrived, I extended my hand to shake his, but Lee politely declined. The damage to his skin from the lymphoma had affected his palms and had caused physical pain, also prompting embarrassment. This reaction is not uncommon among patients with this disease, many of whom feel uneasy in social interactions and sometimes are embarrassed to be seen in public, be touched, or give touch. I have witnessed that in the patients I cared for and was sympathetic to what Lee was likely going through.

"How was your trip?" I inquired, hoping some small talk would put him more at ease.

"Tiring, but OK," he replied.

"When do you head back?" I asked.

"Today, right after we are done."

I was pleased to see that he looked a lot better than I would have expected, especially considering that he'd just flown halfway across the country. As we talked, he answered my open-ended questions about his diagnosis with well-informed answers. Once in a while he got some of the dates mixed up in a minor way, compared to what was in his records, but that didn't surprise me. After all, he had been under a great deal of stress. And a foggy memory is a known side effect of the chemotherapy and radiation treatment he'd received for his cancer.

Lee, as I learned, was a strong guy. He'd always been a hard worker, and a fairly content guy, too. He lived in a rented home

Figure 9: (Top) The tank and equipment used by Lee Johnson. *(Bottom)* Lee Johnson wearing his protective gear while on the job. Both photos were shown as plaintiff's exhibits during the trial.

with a big backyard. "Before I got sick, life was pretty good," he would tell journalist Carey Gillam more than a year later.[2]

Part of his job with the school district, Lee told me, involved pest management. He caught skunks, mice, and raccoons, and patched up holes in walls in order to keep them out. He also worked on irrigation and sprayed pesticides. He frequently used Ranger Pro, an herbicide marketed by Monsanto that has glyphosate as its main ingredient. In his November 2018 interview with journalist Carey Gillam, he explained that in the mornings he would fill up a container with pesticide concentrate and put that in the back of his truck. He would then mix up a tankful of spray, what he called "the juice," and started spraying (see figure 9). Though he didn't like working with chemicals, Lee enjoyed the rest of his job.[3]

"I really loved my job," he told me.

"What did you like about it?" I asked.

"It's a steady job. I felt I was helping. It allowed me to provide and care for my family. I know it might sound crazy to some people, but I was delighted to go to work every day."

Lee went on to tell me how he exercised caution when using Ranger Pro and how he followed all recommendations that were provided to him.

"Did you ever get training on that compound?" I asked.

"I did when I joined," he said. "I had to attend a class or some course. We were all told it is so safe and it won't cause any problems. I remember someone telling me that this stuff is so safe we can drink it."

I could hardly believe what I was hearing. "What?"

"That's right," he said, making eye contact with me for the first time. "They told us it was safe enough to *drink*."

I shook my head in disbelief.

Lee went on, getting visibly angry as he continued. "I mean, I did everything I could, didn't I?" he said. "I wore protective equipment, I trained, I followed the rules, and I ended up here. No one told me that this thing could cause cancer."

He went on to tell me how, despite his precautions, his skin would come in contact with the herbicide, and when the weather was windy, it sometimes got on his face. He also shared with me how once when he was spraying, a hose broke, and he was saturated with Ranger Pro—even under the protective clothing he wore.

Lee began spraying in 2012, and he started developing an unusual rash sometime in the spring of 2014. At first his doctor wasn't sure what it was; he had to have skin biopsies done and analyzed by another institution that had more expertise in such skin malignancies. When Lee received the diagnosis of lymphoma, he decided to call Monsanto to see if it could be related to the spraying.

Message

From:	GOLDSTEIN, DANIEL A [AG/1000] [/O=MONSANTO/OU=NA-1000-01/CN=RECIPIENTS/CN=527246]
Sent:	11/11/2014 8:19:51 PM
To:	BIEHL, PATRICIA M [AG-Contractor/1045] [/O=MONSANTO/OU=NA-1000-01/cn=Recipients/cn=208718]
Subject:	RE: Ranger Pro Exposure

I will call him.   The story is not making any sense to me at all.

Dan

From: BIEHL, PATRICIA M [AG-Contractor/1045]
Sent: Tuesday, November 11, 2014 2:12 PM
To: GOLDSTEIN, DANIEL A [AG/1000]
Subject: Ranger Pro Exposure

Spoke with Dewayne Johnson @ ▮▮▮▮▮▮ and this is his story.

He told me he works for a school district in CA and about 9 months ago had a hose break on a large tank sprayer. This resulted in him becoming soaked to the skin on his face, neck and head with Ranger Pro. He said he was wearing a white exposure suit and it even went inside that. A few months after this incident he noticed a rash on his knee then on his face and later on the side of his head. He said he changed his laundry detergent, dryer sheets and used all creams available to him but nothing seemed to help. His entire body is covered in this now and doctors are saying it is skin cancer.

He is just trying to find out if it could all be related to such a large exposure to Ranger Pro since he stated his skin was always perfect until this happened. He is looking for answers.

Thanks in advance for your assistance.

Patricia Biehl
Product Support Specialist
314-▮▮▮▮▮▮
800-768-6387

Figure 10: Email from Dr. Goldstein when he was informed about Lee Johnson's condition.

"I asked them if what I am having was related to that juice," he told me.

"What did they say?"

"They never got back to me," he replied bitterly. "No one called me back."

Later, I learned through public records and discovery of internal email communications that one of Monsanto's medical directors did in fact hear about the call, but he never got back in touch with Lee. What's more, he seemed to doubt Lee's account. "The story is making no sense," Dr. Daniel Goldstein wrote in an email (see figure 10).

Lee had done the right thing by calling the manufacturer. I thought that, regardless of Dr. Goldstein's opinion, a call back by someone from Monsanto would have been not only Customer Service 101 but, in this case, the logical and most humane thing to do. To me, Monsanto's silence spoke volumes.

I learned a lot from Lee, just as I have learned from all the cancer patients I have cared for over the years. It is never easy to face your own mortality. I saw the courage in Lee's eyes, but also the disappointment that somehow the system had failed him. He did everything by the book and trusted those who were supposed to protect him from preventable ailments.

Lee had a form of non-Hodgkin lymphoma called mycosis fungoides, which is a T-cell lymphoma that affects the skin and sometimes internal organs as well. Many patients with this disease can be mistakenly thought to have eczema before the correct diagnosis is made. While most patients have a slowly progressing disease that they can live with for many years, it sometimes transforms into a more aggressive illness. This was the case for Lee. In September 2016, he had additional biopsies done that confirmed a transformation of his lymphoma into a more aggressive form with an average survival rate of two years. Suddenly this was no longer just a skin disease that caused him discomfort and embarrassment. This was now a matter of life and death.

Lee's records showed that he had been seen at various prestigious medical institutions in the Bay Area, and in my opinion his care had been both adequate and appropriate. At the time I saw him, he had lost his insurance and was waiting for state coverage to kick in. In fact, he was supposed to have additional imaging done so that his doctors could determine next steps, but that couldn't happen until his coverage was approved.

I had Lee lift his shirt up to listen to his chest and heart. His skin was mottled by angry red and bluish welts (figure 11). After a brief examination where I inspected these lesions and ulcerations, listened to his heart and lungs, and felt for enlarged lymph nodes, we were ready to part ways.

He thanked me, though once again he was embarrassed to shake my hand as he got ready to leave.

"Good luck with the test results, Lee," I said. Because I felt he needed a little encouragement, I told him that I thought he was receiving excellent care from his physicians. "I'm sure they're doing their best, and that they're going to try every option to get you better."

He nodded and smiled tersely. "Thanks, doc."

That night, I returned home and held my twin boys tightly, squeezing them and not wanting to let go. I could only imagine the pain that Lee was going through, and my heart was pierced by the thought that, given the sensitivity of his skin, Lee probably could no longer hold his children the way I held mine.

Four months after meeting Lee, on January 15, 2018, I was deposed again by the Monsanto attorneys. A new article had just been published in the *Journal of the National Cancer Institute* (*JNCI*). It was an update on an ambitious, ongoing investigation called the Agricultural Health Study (AHS).[4] Because the article

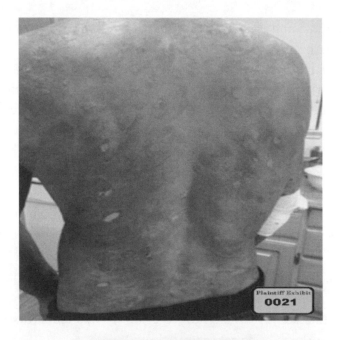

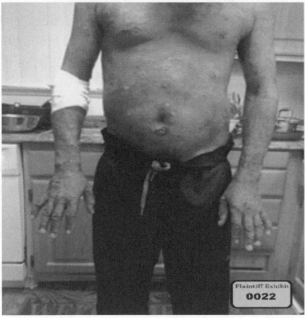

Figure 11: Lee Johnson's lymphoma. These pictures were shown as plaintiff's exhibits during the trial.

had just come out online and was certainly going to be a topic of discussion during the trial, the Monsanto lawyers were interested in my opinion about it.

The AHS was what's known as a prospective cohort study, a type of research project that follows a group of people over time who are similar in many ways but differ in terms of a particular characteristic and compares them in terms of a particular outcome. The AHS enrolled more than 50,000 individuals applying for a pesticide license in Iowa and North Carolina between 1993 and 1997. These applicants (and their spouses, where applicable) had to complete a questionnaire about their pesticide exposure going back as far as twenty years. The applicants were asked about all pesticides they might have used (including glyphosate) and whether they wore protective clothing while using the substances. The idea behind the study was to examine the development of various cancers in this cohort by monitoring the cancer registry (a system for the collection, storage, and management of data on persons with cancer) in both states. By looking at the incidence of various cancers, investigators could then study whether the applicants' pesticide exposure was a factor in their illness. One of the cancers being looked at was non-Hodgkin lymphoma.

One major problem with this study was the fact that, as we've seen, glyphosate use increased significantly after 1995, but the study assessed exposure in terms of the date of a person's license application, and many people started using glyphosate after they were initially surveyed. For instance, if an applicator who applied for a license in 1993 said on the questionnaire that they had never used glyphosate, that person would be categorized as unexposed. Let's say that later on this person started using glyphosate and eventually went on to develop lymphoma. This study would still classify the patient as unexposed.

The AHS investigators realized this flaw, and in 1999 they decided to send a second questionnaire to the original applicants in order to explore changes in their patterns of pesticide use. But that second survey only asked about pesticide use during the previous year, not any further back. Furthermore, between 37 and 38 percent of the original survey respondents never answered this second questionnaire, so their data was missing. The authors of the *JNCI* article utilized a mathematical model called imputation, which essentially attempts to predict what the answers would have been from missing applicants. I would call this a guess—an educated guess, maybe, but conjecture, nonetheless. Why someone might have dropped out of the study was unknown to anyone but the person who dropped out, and in my humble opinion, no mathematical model can ever rectify this problem with a high degree of certainty.

You don't need to be a scientist to recognize the flaws in the AHS. However, because it concluded that there was no relationship between glyphosate and lymphoma, this became Monsanto's favorite study and the one that they would bring up again and again in all of the trials.

Interestingly, Monsanto's scientists themselves had criticized this same study years before it was published, because they were nervous it might show a link between glyphosate and non-Hodgkin lymphoma. But now that it showed no danger associated with Roundup, it was being hailed by Monsanto as a landmark in the history of cancer research.

In my January 2018 deposition, I pointed out the various aditional weaknesses of the AHS. For example, the control arm (those who were not exposed to glyphosate) consisted of pesticide applicators and farmers, both groups with a well-established risk of developing non-Hodgkin lymphoma. When the control group is at elevated risk, establishing higher risk in the exposed group becomes more challenging in my opinion. Furthermore, I suggested that a mathematical model applied in a scenario with an almost 40 percent dropout rate and missing data was of little value to me when counseling patients. But my opinion appeared to fall on deaf ears during the deposition; nothing surprising there.

Just as with the first deposition, I had to pay attention to how Monsanto's lawyers phrased each question, knowing that they were trying to get me to say something that would make me look bad. This time they introduced a new strategy: the "double negative." That is, they would phrase a question using two negative words (imagine questions like "Were you not unhappy about the result?" and "Were you not surprised by a negative study?"), and I would have to listen ever so carefully to even understand what they were asking. To the credit of the Monsanto lawyers, they did repeat the questions when I requested it, but on several occasions, I suspected that the ambiguity was intentional, so that they could have favorable sound bites to display to the jury when I appeared in the courtroom during the trial.

I was being deposed again by Kirby Griffis, the same lawyer who had deposed me back in the summer of 2017. We spent a lot of time discussing the new *JNCI* article, as this memorable exchange from the transcript shows:

Kirby Griffis: *You would agree that this is a piece of evidence that weighs against causation; correct?*

Chadi Nabhan: *It is a piece of evidence that suggests no causation between glyphosate and non-Hodgkin lymphoma. Which I don't agree with.*

KG: *So, you agree that it is a piece of evidence against causation, but you disagree overall with that conclusion that there is no causation; is that accurate?*

CN: *I do disagree with the conclusion, yes.*

KG: *And the rest of what I said is accurate as well; correct?*

CN: *Yes.*

Some of the data from the AHS had been published before, notably in a 2005 article that presented similar findings.[5] In my opinion, the *JNCI* article included a longer follow-up and some additional data, but those elements did not overcome the problem of missing information in the AHS, as mentioned earlier.

Between the first deposition and this second one, I had reviewed more literature, animal studies, toxicology data, and other information, all of which had solidified my opinion regarding how glyphosate could cause lymphoma. Griffis seemed to want to use that against me. "What size new epidemiology study would it take to shake your conviction?" he asked.

"The body of evidence so far that I have reviewed is convincing that there is a causation and an association between glyphosate and non-Hodgkin lymphoma," I said. I went on to explain that the *JNCI* article was an update of that 2005 article, and that I had taken the earlier article into full consideration.

He later asked me once again if I agreed that the conclusions of the *JNCI* article were supported by the evidence that was provided in the article. I answered: "The authors' conclusions are supported by the evidence that they actually showed. The evidence has a lot of flaws, and subsequently the conclusions will have a lot of problems. But, yes, their conclusions are supported by the evidence that they evaluated."

Two weeks after that deposition, on January 30, 2018, I was deposed yet again, this time in Lee Johnson's specific case against Monsanto, where I would be testifying as an expert witness for the plaintiff. This time they wanted to determine why I thought Lee Johnson's lymphoma was caused by his exposure to the weed killer.

For this deposition, too, I would be questioned by Kirby Griffis. "We meet again," I said cheerfully as I strode into the room and saw him, as impeccably dressed as ever. "I feel like we're old friends." All I got in return was a polite smile.

I was also familiar with his line of questioning by now. He started with seemingly innocuous, even irrelevant-sounding questions about invoices, hospital privileges, and whether I was seeing patients. He followed these with some general questions about non-Hodgkin lymphoma and how many patients with Lee's form of the disease, mycosis fungoides, I might have seen when I was in clinical practice.

Griffis's tone changed when I said I was unable to accurately estimate Lee's prognosis. They wanted to challenge my original assessment that he had about two years to live from September 2016, which was when his disease, originally diagnosed in 2014, took a more aggressive form.

KG: You said that the life expectancy—Mr. Johnson's life expectancy was approximately two years, correct?

CN: Based on the statistics, that's what I projected.

KG: And that's two years from the large-cell transformation being diagnosed, correct?

CN: Generally speaking, that's the median. As you and I know, median means some patients will exceed the median and some patients will fall short of the median. I have lost patients in less than twelve months after large-cell transformation, but based on the information provided, my projection here would have been twenty-four months as a median survival time.

KG: So, the median survival time you were projecting would have been September of 2018; right?

CN: Something like that.

KG: And you note in the next sentence Mr. Johnson is nearing the end of that period. He was just a few months away from it at that time?

CN: That was before I met him.

KG: And his disease will continue to progress and show resistance to treatment, correct?

CN: No doubt in my mind.

KG: When you met him, did that extend your estimate of his life
expectancy?

CN: I was pleasantly surprised to see that he was doing better than
I expected, but I still believe he has limited life expectancy. His
skin disease is going to progress rapidly. I don't believe he's going
to respond to treatments very well. He might respond temporarily.
The duration of response will be short. And I would be shocked if
he's around next year.

Clearly, I was proven wrong on that point, for I am happy to say that Lee is still alive at the time of this writing. I am ecstatic that my projections about Lee were wrong. And it's not uncommon for physicians to venture projections on survivability that turn out to be either overly optimistic or overly pessimistic. But patients and their families want answers, and "How long do I have to live?" is certainly one of the most basic questions they would ask.

Griffis went on to ask many questions about other factors that might have played a role in Lee's developing mycosis fungoides. He asked my opinion on whether tobacco and alcohol could cause non-Hodgkin lymphoma, and I answered that I didn't think they were associated with that illness. The Monsanto team was also focused on the chronology of events, and specifically whether Johnson had been exposed long enough before he developed the lymphoma (something known as "latency period," the time from when a human is exposed to a known carcinogen until the time the actual cancer is discovered). Monsanto was arguing that because Lee first became employed at this job in June 2012 and a rash was initially found in September 2013, the latency period was not long enough to suggest a relation between glyphosate exposure and non-Hodgkin lymphoma. But I countered that this was not true. His medical records did not definitively confirm a rash in September 2013, but did confirm one between April and June 2014, by which point Lee would have been continuously exposed to this hazardous substance for two years. I also explained to the attorneys during my deposition that the latency period for non-Hodgkin lymphoma varies, and there is no minimum or maximum latency period before someone develops the disease. I further reminded the Monsanto lawyers that Lee's exposure history was very heavy. Spraying was part of his daily routine on the job. Furthermore, he had had several accidents in which Roundup was spilled on him.

Griffis then asked me a couple of hypothetical questions. "There are lots of patients that you have seen with mycosis fungoides for whom you have no idea what caused it, right?" he said.

"That is true," I replied.

"Mr. Johnson could well be someone who would have developed mycosis fungoides when he did whether he was exposed to glyphosate or not for all you know, correct?"

"He could have," I answered.

A seemingly innocuous exchange, but one that would play an important part in the courtroom a few months later.

At the end of this third deposition, I was once again exhausted. I was anticipating becoming involved in more cases as Tim had told me that he would introduce me to other attorneys that might need my assistance. I was starting to think how challenging it would be to find dates for other plaintiff depositions. And I was finding it difficult to find the time to fit in all the work that was needed. Meanwhile, I was becoming a bit more irritable at home because I was struggling to balance my travel schedule, work on the litigation, and family duties. But I was committed to seeing this through. The more I learned about Monsanto's tactics and the more I learned about Lee Johnson, the more determined I was to do everything I could to help get justice for this man.

5

The Night before the Daubert Hearing

My first appearance in court was not in front of a jury but rather before a judge: the Honorable Vince Chhabria (see figure 12).

I had previously seen his name on some of the documents and court orders that had been issued in the Johnson case over the preceding couple of years, and I was curious to learn more about him, so I looked him up. I learned that Chhabria had been born in 1969 in San Francisco, and that he had an impressive resume, including having worked for several top law firms in the Bay Area before being named deputy city attorney for government litigation in San Francisco in 2005 and co-chief of litigation in 2011, posts he held until 2014, when President Barack Obama nominated him to serve as a judge on the United States District Court for the Northern District of California.

In March 2018 the Miller Firm informed me that I needed to appear in front of Chhabria in San Francisco for something called a Daubert hearing.

"Who's Daubert?" I asked. I thought I'd learned the names of those people involved in the Johnson case, but I didn't remember seeing that one.

Figure 12: Judge Chhabria.

The short answer was that there was no one named Daubert who was involved in this case. Rather, at a Daubert hearing a judge rules on the admissibility of expert witness testimonies in an ongoing litigation case in a proceeding that takes place before the trial, so there is no jury present. This type of hearing is named after a United States Supreme Court case involving birth defects allegedly caused by a woman's use of an anti-nausea drug during her pregnancy. In a Daubert hearing, both sides bring in their respective expert witnesses, who are examined and cross-examined in front of the judge. The judge is tasked with ensuring that an expert's testimony is relevant to the matter at hand and that it rests on a scientifically reliable foundation.

I also learned that while Judge Chhabria would not personally be trying Lee Johnson's case, he was overseeing all of the Monsanto-related litigation that was now beginning to emerge, including Lee's. To that end, in March 2017, Judge Chhabria made a ruling on a key aspect of the overall question of whether Monsanto attempted to influence the scientific research and policy toward glyphosate. To put it very generally, Monsanto was reluctant to disclose some documents that might sway the jury against them, and the plaintiffs wanted these documents revealed because they would work in their favor. The specific documents at issue here were materials submitted in connection with the motion to compel Jess Rowland's deposition. As we saw earlier, Rowland was the EPA official who said he deserved a gold medal for blocking an initiative by the Department of Health and Human Services to investigate glyphosate's association with cancer. Monsanto wanted the judge to seal all but one of the documents connected with this motion, meaning that the jury would not be able to see them. The jury eventually was never shown Rowland's deposition. In my opinion, Rowland was a key person in this case, and I was stunned to learn that Monsanto and the plaintiffs' attorneys were going back and forth as to what documents should and could be shown to the jury. Shouldn't the jury see all documents related to the case they are being asked to render a verdict on? Shouldn't they know all the facts? But no, the jury sees only what the judge deems appropriate. And thus, a lot of energy is spent by both sides trying to convince the judge to allow or disallow certain documents or facts into evidence. Strategy, I came to understand, sometimes trumps facts.

I also learned that the litigation I was part of was now officially called MDL-2741, where MDL stands for "multidistrict litigation." When cases are being brought in several federal district courts on matters that involve common issues (as can often be the case with complex product liability lawsuits, like the ones involving Monsanto and glyphosate), the pretrial proceedings for all of them may be temporarily consolidated and heard before a single court to reduce the burden on the court system and make the proceedings more convenient for all of the parties involved in the lawsuits.[1] Judge Chhabria was presiding over this MDL, which included the hearing I was about to participate in.

All of this was confusing to me, an outsider to the legal system. And the cases I would eventually wind up participating in as an expert witness were connected to the federal proceedings in different ways. Though the attorneys would always explain the intricacies to me, I found it tough to keep track. But I did find out that all the federal cases against Monsanto across the country were to be consolidated into the Northern District of California. Lee Johnson's specific lawsuit was a state case and would be tried in state court in California, but some other plaintiffs' cases wound up being kicked into the federal court system.

I arrived in San Francisco the night before the Daubert hearing and had dinner with lawyers from the Miller Firm at an Italian restaurant in the Union Square section of the city. Tim was absent that evening, working on other aspects of the case. But I met his boss, Mike Miller, the lead attorney of the firm, for the first time. Mike was accompanied by his wife, Nancy Guy Armstrong Miller, herself an attorney, and another lawyer from the firm, Jeff Travers. The four of us sat down at a corner table. I was hungry, but anxiety over my appearance before a judge the following morning made every item on the menu seem unappetizing. I ended up ordering a small dinner salad and plain grilled salmon.

Mike Miller's jovial, outgoing personality helped take my mind off my worries. He had a reputation as a commanding courtroom presence, someone that juries liked. As I picked at my salad, I could see why. I had wanted to learn more about what would happen the following day and get some advice and coaching tips. Instead, we talked about everything else *but* the upcoming Daubert hearing. I recall mentioning to Mike that night that I had just finished reading John Grisham's book *The King of Torts*, which describes the rise and fall of a young tort lawyer and examines the world of attorneys in that legal specialty, including a master of the art who regales the book's young protagonist on the fine points of private jets and heartily encourages him to purchase one with the vast sums he's raking in on these cases.

"Chadi!" he burst out. "That book is about me!" Then he threw his head back and roared with laughter. Jeff did not comment; he was reviewing a couple of papers while at dinner. (I came to learn how brilliant Jeff was despite his quiet demeanor. He knew the literature like no other, and eventually became my go-to legal guy when discussing glyphosate issues.) Nancy shook her head and smiled. She must have heard her husband make this claim before, as if John Grisham had been following Mike around with a notebook for a year. No matter—it's impossible not to like Mike the minute you meet him. His pleasant southern accent and gracious manners certainly don't hurt. Of course, you might assess him differently if, like Monsanto, you were about to face him in a trial. While he was popular with juries, he had a reputation of being fearless and aggressive with his legal adversaries.

I also learned that taking on Monsanto is not without significant financial risks for the legal firms involved. The firms that have joined forces to fight Monsanto in the courtroom need to cover all the costs of bringing these lawsuits, including paying the attorneys and staff, paying all the experts for their time, paying for recordings and court reporters, covering travel costs, and other elements. All this amounts to millions of dollars that the firms won't recover unless the plaintiffs win. The patients who are suing Monsanto assume none of these costs, and of course receive nothing unless the cases are won, but if an award is granted, they share some of it with their lawyers. Plaintiffs' lawyers understand these risks, and that's why they typically take on cases like these only if they strongly believe they might win. That's also why, generally, several law firms join forces, so that the costs are distributed among them.

When dessert and coffee were ordered, we finally got around to the Monsanto case and what to expect the following morning. I told my dinner companions that I had never testified as an expert witness in a courtroom before and was getting more nervous as we got closer to the actual trial. Mike smiled reassuringly. "Just remember that you know more about lymphoma than anyone else in the courtroom," he said.

While I would remember those words, and they would prove helpful to me not just the next day but in the months ahead, I still stayed up almost all night preparing for the Daubert hearing. I reviewed the literature that I had started working on after my first phone call with the Miller Firm, now almost two years in the past. At this point, I felt like I could teach a seminar on glyphosate. But this was a courtroom I would be heading into, not a classroom. I carefully organized my thoughts about how I had reached my conclusions.

Among the most important items to review were the epidemiological studies that I relied upon, which reached different conclusions than the Agricultural Health Study had. I knew I'd be grilled on these studies. I was always on the lookout for more evidence about glyphosate, and I'd found it in an unlikely place: Sweden. The Swedes have maintained an excellent national cancer registry. Most countries have them, but few such registries are as efficient as the Swedes'. Established in 1958, it covers the country's entire population of about 10 million. Approximately 50,000 cases of malignant cancers are registered in Sweden every year, and it is mandatory for every physician to report newly diagnosed cases to the registry, providing a rich trove of information for those investigating epidemiological questions. The data available to researchers includes patients' sex, age, place of residence, site of tumor, type of cancer, stage of disease, and date of diagnosis. Notably, there is also follow-up data on the date and cause of death, and multiple checks and balances in place to minimize errors in reporting to the Swedish registry.

For the purposes of my research, the rigor of the Swedish cancer registry was a godsend, as several important epidemiological studies, using the information from the registry, have been conducted since 1999, when Dr. Lennart Hardell first suggested a link between glyphosate exposure and the development of non-Hodgkin lymphoma.[2] Three years later, Dr. Hardell published another study showing a similar link, although the finding was not statistically significant.[3] Essentially, statisticians argue that if a study is statistically significant, then it is meaningful and reliable more than 95 percent of the time. Still, to me, the evidence was mounting up.

In 2001, Canadian investigators led by H. H. McDuffie published a paper demonstrating that the more exposure one has to glyphosate, the higher the risk of developing non-Hodgkin lymphoma.[4] The Canadian investigators pulled the lymphoma cases from the Canadian national cancer registry and attempted to interview every patient, asking about their exposure to various types of pesticides. They then asked the same questions of the control patients, those who didn't have lymphoma. The researchers found that the risk of developing non-Hodgkin lymphoma was slightly more than doubled among individuals who were exposed to glyphosate more than two days per year.

I knew that Monsanto would push back against the findings of this 2001 study because they had already asked me about it during my August 2017 and January 2018 depositions. Essentially, Monsanto was arguing that the researchers had not accounted for possible exposure to the other pesticides that might have contributed to the development of non-Hodgkin lymphoma. (The technical term for this is "confounding"—that is, there may be another, unknown factor that affects a potential cause-and-effect relationship.) Monsanto's goal was to implicate other agents as possibly having caused the lymphoma—but not theirs.

Unsealed documents later revealed that Monsanto appeared nervous about the findings of the McDuffie study. I can imagine they were very satisfied *not* to see the word "glyphosate" mentioned in the study's abstract. This, wrote Monsanto scientist Donna Farmer in an email exchange with one of her colleagues, "is a huge step forward—it removes it from being picked up by abstract searches" (figure 13) and thus made it less likely to be found by researchers looking for evidence of a link between the

Message

From:	FARMER, DONNA R [AG/1000] [/O=MONSANTO/OU=NA-1000-01/CN=RECIPIENTS/CN=180070]
Sent:	11/29/2001 2:07:23 PM
To:	ACQUAVELLA, JOHN F [AG/1000] [/O=MONSANTO/OU=NA-1000-01/CN=RECIPIENTS/CN=145465]
CC:	GOLDSTEIN, DANIEL A [AG/1000] [/O=MONSANTO/OU=NA-1000-01/CN=RECIPIENTS/CN=527246]; ARMSTRONG, JANICE M [AG/1000] [/O=MONSANTO/OU=NA-1000-01/CN=RECIPIENTS/CN=597137]; HEYDENS, WILLIAM F [AG/1000] [/O=MONSANTO/OU=NA-1000-01/CN=RECIPIENTS/CN=230737]
Subject:	RE: the McDuffee article appears - glyphosate not mentioned in the abstract

John,

I know we don't know yet what is says in the "small print" - but the fact that glyphosate is no longer mentioned in the abstract is a huge step forward - it removes it from being picked up by abstract searches!

Donna

-----Original Message-----

From:	ACQUAVELLA, JOHN F [AG/1000]
Sent:	Thursday, November 29, 2001 7:54 AM
To:	FARMER, DONNA R [AG/1000]
Cc:	GOLDSTEIN, DANIEL A [AG/1000]; ARMSTRONG, JANICE M [AG/1000]; HEYDENS, WILLIAM F [AG/1000]
Subject:	the McDuffee article appears - glyphosate not mentioned in the abstract
Importance:	High

The McDuffee article appeared in the November issue of the journal Cancer Epidemiology, Biomarkers, and Prevention (see abstract below). Unlike the abstract presented at the International Society for Environmental Epidemiology meeting August 1999, Glyphosate is no longer mentioned as a risk factor in the abstract. I'll have to get the article and see what it says in "the small print."

John

Figure 13: Donna Farmer's email showing her satisfaction that glyphosate is not mentioned in the abstract of the McDuffie study.

chemical and non-Hodgkin lymphoma. Dr. Daniel Goldstein (the doctor who had questioned the veracity of Lee Johnson's claims) was copied on that email. When asked during one of his depositions why Monsanto would be celebrating lack of easy access to research about glyphosate, he answered that he could not read Farmer's mind and so could not say what she had meant by that.

Beyond the Canadian study, another study from Sweden confirmed that the more exposure to glyphosate a person has, the higher the risk is of developing non-Hodgkin lymphoma.[5] I thought this was a significant finding. This 2008 study, led by cancer researcher Mikael Eriksson of Sweden's Lund University Hospital, showed that being exposed to glyphosate for more than ten days over the course of a lifetime more than doubled the risk of non-Hodgkin lymphoma.

What was beginning to seem clear was that there was a dose response to glyphosate. In other words, the more you were exposed to it, the greater your chances appeared to be of developing non-Hodgkin lymphoma. This would strengthen the case for someone like Lee Johnson, who had used it regularly as part of his job for years. So, it was no surprise, then, that Monsanto tried to downplay the findings of the Swedish study along with the Canadian study.

Monsanto also challenged the findings of an American study published in 2003 by Drexel University investigator A. J. DeRoos and colleagues.[6] In it, glyphosate was found to double the risk of non-Hodgkin lymphoma even after adjusting the results to account for more than forty other pesticides.

Even as I processed the statistics, I also tried to bring to bear my background as a clinician—a physician who works with the people behind these numbers, making decisions about their care. Throughout my experience with the Monsanto trials, I always felt that I needed to exercise my clinical judgment, in addition to assessing the statistics, to determine the reliability of any given study, because the factors that go into treating any given patient are even more complex than the formulae used to crunch the numbers in a study.

I hoped that my perspective on the disease and my experience dealing with the flesh-and-blood embodiments of these statistics—the human beings and their families who were suffering from a terrible disease—would be part of my value in helping make the case for Lee Johnson and the others.

This, I believed, was part of what Mike had been trying to tell me with his encouraging words. But still, like most people who aren't lawyers, I had no idea what to expect in a courtroom, much less a federal courtroom. That was why I stayed up late in the hotel room going over the research, my prior depositions, my expert report (which contained the analysis of the many studies I had found), and other relevant materials. Finally, I tumbled into bed, but, unable to sleep, I thought about what would transpire the next day. I was about to meet Judge Chhabria for the first time. The trial would begin soon after. I had heard that reporters from major news outlets around the country might be present. The potential size and scope of this litigation were enormous, so the entire legal world was going to be watching these proceeding with interest as well—watching proceedings in which I would play a small but maybe significant role.

How had I gotten to this point? How had I, a Syrian immigrant, a physician who originally had aspirations of becoming a journalist, end up as a medical witness in what I now realized was going to become one of the largest and most important lawsuits in American history? That long, tortuous route, I realized, may have started with a conversation with my father many years earlier. As I stared at the ceiling of my room in the Westin, I traveled back in time, to my childhood home in Damascus.

6

Coming to America

My normally mild-mannered father was appalled.

I had recently graduated from high school in my hometown, Damascus, Syria, and my grades were good enough to allow me to apply and likely enter medical school—something my dad had always dreamed of. Instead, I had to break the news to him that my real calling was journalism.

I had grown up fascinated with the news media. I wanted to be a part of a profession in which ideas mattered and writers articulated their opinions in powerful prose on the op-ed pages. I wanted to be a member of a profession that spoke truth to power, as they say now, and whose work could in some cases compel new laws or force the resignation of politicians.

My dad held no such lofty views. "A journalist?" he said, as if wounded in the heart. "Why?"

I explained to him that this was a career in which I could have an impact, and one where I could work with words. "Dad, you know I love writing," I reminded him. "I even wrote poetry. I think I'd be good at it."

"You do know that your grades will allow you to get into any medical school in the country, don't you?"

Considering that there were only three medical schools in Syria at the time, I found that distinctly underwhelming.

"I know that, Dad," I said. "But I'm not sure. I think I have other talents I'd like to explore."

My dad was an engineer, however, and his reasoning was as clear and well-integrated as a circuit board.

"Son," he replied, "someday you will have a family. The medical profession will provide security for that family. You'll have a financial cushion; you will worry less about money . . . and you will be helping people!" Seeing that those words hadn't convinced me, he continued, "Please think carefully, Chadi. You can always write on the side and explore your other passions while you're practicing medicine."

"I'll think about it," I said, my expression and tone clearly indicating that I wouldn't.

He knew how stubborn I was. At that point, he went from reasonable dad to "I've had enough of this" dad. (As a father myself now, I recognize the occasional need for the latter.) "You will apply to medical school," he said, his tone making it clear that this was non-negotiable.

Six years later, a newly minted MD, I stood in line at the US embassy, located on a tree-lined street in one of Damascus's most prestigious neighborhoods. There were about a dozen other people waiting as well. Some were applying for tourist visas, others for immigration, and many, like me, for an F-1 student visa. Each of us had a brown folder filled with documentation, applications, and affidavits. Mine held my medical school transcripts, diploma, letters of recommendation from some of my professors, and bank documents showing that I could support myself in the United States, not to mention a multi-page application that had been daunting to fill out.

In my head, I rehearsed the answers to the questions I had been told to expect from the US consul who would review my application. For example, some of my friends had cautioned me that I should never, *ever* say that I planned on remaining permanently in America. If I did, they said, I wouldn't be granted the visa.

"Then what should I say?" I asked them.

"Just don't say that!" they said.

Okay, fine. Mum's the word. If I'd been told that I should memorize "The Star-Spangled Banner" and sing it out loud, I would have done that. Anything to help me get into the United States.

Why America? My parents lived comfortably in Syria. Despite what you may have heard about it—and certainly there have been sad and terrible events there in the past decade—Syria is a beautiful country with a fascinating, rich, and ancient culture. At the time, it was also a country in which there was little unemployment and most citizens had a fairly high standard of living, especially compared to some other parts of the Middle East. Moreover, it was also a secular nation, so we enjoyed many of the same basic freedoms as Americans or Europeans.

But it was not America. That was where many young Syrians wanted to be at the time. Even from nearly seven thousand miles away, America seemed big, glamorous, exciting, and ripe with opportunity and possibility.

At last, my name was called, and I walked over to one of the windows. A middle-aged man with gray sideburns looked at me from the other side of the glass. He extended his hand, indicating that I should slide my papers beneath the glass that separated us.

"So, why are you going to Boston?" he asked as he looked through the papers.

"I will be studying and preparing for the USMLE examination to start residency," I said in my carefully rehearsed English. I didn't have to tell him that this stood for United States Medical Licensing Exam; I'm sure many others had stood before this same consular official with the same request.

"Okay, but why Boston?" he asked, as if I hadn't understood his question in the first place.

"A . . . a friend of my family lives in Boston," I stammered, unprepared for that particular question. "And my family thought he could help me find a place and get situated in a safe area."

"Are you saying there are unsafe areas in the United States of America?"

I could see my F-1 visa flying out the window of the embassy into the clear-blue Syrian sky. I had screwed up. I had just criticized the country that I wanted to study in—and to a member of that country's diplomatic corps, to boot!

I thought quickly.

"No, sir, I'm not saying that at all," I said apologetically. "It's the movies I sometimes watch that portray that."

The consul smiled warmly. "I've seen some of those movies, too," he said. "And if I wasn't from the U.S., I'd think we were all running around like Terminators trying to shoot each other."

I stood nervously as he thumbed through my paperwork and asked a few further questions about my background and my studies. Then he folded everything up and looked at me with sharp eyes. "Are you planning on coming back here after you finish your studies?"

Honestly, I had no idea. It was early August 1992, and I was a few weeks shy of my twenty-fourth birthday. At that age, who knows what the future holds? This was going to be the greatest adventure of my life. I remembered what my friends had said, however. *Don't tell them you might not come back.* Instead, I made an on-the-spot decision to be honest—something that has remained a guiding principle for me in life. "I am not sure, sir," I said earnestly. "It's hard for me to predict. I can't be 100 percent certain."

He appeared to have concluded reviewing all my documentation. A slight grin worked its way across his face, and he looked me straight in the eye. "Leave your passport here and we will stamp the visa after a few procedural issues," he said. "Come back after 2 p.m. and it will be ready."

I tried to stay cool and hide my excitement as I thanked him. Then I exited through the heavily guarded gate, walked a few blocks back to one of Damascus's commercial streets, and hailed a taxi. I couldn't wait to get home to share the good news with my parents, other family members, and friends.

My mom cried when I told her. I wasn't sure if she wept because thousands of miles would now separate us for the first time since I was born or because she was happy for me; likely it was a combination of both. My mom and I have always been close, and the thought of being far away from her and not seeing her on a regular basis made my heart ache and my eyes fill with tears.

My friends soon arrived at the house to join the celebration. They took full credit for my success at the embassy. "Good thing you followed our advice," they said.

My dad just smiled. "I'm very proud of you, son."

There wasn't anything special about me or my application. I was just lucky that day. And although I have returned to Syria many times to visit, I still feel lucky to be here in the United States thirty years later.

I had heard Boston was cold. But cold in Syria is anything less than 50 degrees Fahrenheit. No one had warned me about the frigid blasts of wind blowing off the Charles River, or temperatures that could plummet to single digits. To counter what seemed to me like polar conditions that first winter, I blasted the heat in my apartment 24/7. One night the landlord banged loudly on my door. "Look at these heating bills!" he shouted, waving some papers at me. "You're costing me a fortune." I promised to turn down the heat and put on an extra sweater.

The climate wasn't the only thing I had to adjust to in this new society. I had to learn how to use the T, as Bostonians call their mass transit system, so that I could commute from my tiny third-floor rental in the working-class suburb of Quincy, Massachusetts, to my studies in downtown Boston.

There, at the Kaplan Educational Center on Boylston Street, I worked on my English and prepared for the USMLE examinations. I relied on TV shows to supplement my language lessons and get acclimated to the culture. *Seinfeld* came in handy every Thursday evening. For a show supposedly "about nothing," it certainly meant a lot to me. It was a glimpse into the American culture that I was learning, and while I'm happy to say that today I don't sound like Cosmo Kramer or George Costanza, I was both entertained and enlightened by them.

I also made a habit of watching Jay Leno's monologue on the *Tonight Show* because he spoke slowly and clearly enough for me to grasp the words he was saying. But while the audience convulsed in merriment at his one-liners, I usually just sat there, not laughing; I didn't really get American humor at the time. Or maybe it was because, between learning English and preparing for the exam, I was studying sixteen hours a day, and it didn't feel like I had time for jokes.

Despite my serious demeanor and the freezing temperatures, I eventually warmed up to Boston. Indeed, while I would leave the city in 1995 to start my residency at Loyola in Chicago, I still have a soft spot in my heart for Boston, the city that taught me English, showed me American culture, and pointed me in the direction of my career in oncology. Boston also taught me some basic economics—as in how to survive on very little money. I took advantage of every McDonald's deal. When they offered two Big Macs for $2, I grabbed those greasy burgers as fast as I could.

Boston also introduced me to American football. I recall the first time I watched a Patriots game on TV, with no clue about what these big guys were trying to do. Unlike the football I knew, soccer, they had goalposts but no goalkeeper. They would run furiously for a few seconds, but then stop. How could you tell who was who under those helmets and pads? And why do they count every score as six points? American sports were strange, I thought. Eventually, I grew to learn and love the game, and to this day, the Patriots are my favorite team. (And let the record show I wasn't some bandwagon fan: this was years before Tom Brady was drafted and they started winning all those Super Bowls.)

Six months after my arrival in Boston, I decided to reward myself by going to the movies. I went alone, because I didn't have many friends. I knew a few students who were also at the Kaplan Center, and of course there was the family friend who helped when I first arrived, but he was older than me and very busy with his own life, so I didn't want to be a burden. I headed over to Copley Square and bought a ticket to watch *A Few Good Men*, starring Tom Cruise and Jack Nicholson. I thought of it as a sort of test—how much could I comprehend? Though I understood only about half of the dialogue, I thought it was a terrific film. I certainly grasped the drama of the famous courtroom exchange between Nicholson and Cruise, when the latter says, "I want the truth!" and the former replies, "You can't handle the truth!" I walked out of the Copley Square theater feeling like I had passed the test: I was finally semi-conversational in English.

I'm not sure why, but that long-ago afternoon—and the experiences leading up to it—flashed through my mind as I stared at the ceiling of my hotel room at the Westin in San Francisco the night before the Daubert hearing.

You can't handle the truth. Isn't that essentially what we were saying to Monsanto? That they couldn't handle the truth about what the research was showing? That they couldn't handle the truth that while products like Roundup were hugely profitable, they also had the potential to cause cancer?

I wondered if Judge Chhabria would see it that way.

7

Daubert Day

Twenty stories high, the Phillip Burton Federal Building soared into the misty morning sky. I had to tilt my head back to take in the scale of this, the largest federal building in San Francisco—indeed, the largest west of the Mississippi River.

Its grand-sounding address, on Golden Gate Avenue, added to the gravitas of the monolithic structure. Built in 1964 and redesigned in the early 2000s, the building houses a roster of tenants that bespeaks a projection of federal power as sweeping as any you'll find outside of Washington, DC. In addition to one of the four federal district courthouses of the Northern District of California, the Burton Building is also home to the local and regional offices of an alphabet soup of agencies, including the FBI (Federal Bureau of Investigation), DEA (Drug Enforcement Administration), IRS (Internal Revenue Service), ATF (Bureau of Alcohol, Tobacco, Firearms, and Explosives), EEOC (Equal Employment Opportunity Commission), FPS (Federal Protective Service), FOH (Federal Occupational Health), USCIS (US Citizenship and Immigration Services), the Federal Public Defender, and the US Passport Agency. Maybe I missed one—was there an office here for, perhaps, the Bureau of Labor

Statistics? I'm not sure, but I do know that on a typical weekday, like the one in which our hearing was to take place, the Burton Building hummed with activity, as citizens and those charged with representing them went about their business with various manifestations of the US government. Typically, I was told, it took a half hour to get through security, especially in the morning, when lawyers arrived. No private security guards here, either—this place was guarded by elite US marshals, who I mistakenly had thought existed only in Westerns. Honestly, I can't remember how long it took me to get access that day, probably because I was preoccupied with my part in the drama that was about to unfold within these corridors of justice.

Waiting to go through security, I also reflected on the namesake of this building, and the reason that had brought me there that day. An eleven-term congressman from California, Burton referred to himself as a "fighting liberal." When he died in 1983 of a heart attack at age fifty-six, former vice president Walter Mondale called him "a warm and caring man, a gifted servant of the people of California, a fighter for social justice, and a national leader in the Congress."[1] That didn't sound to me like someone who would condone a giant chemical corporation challenging any criticism about a product of theirs that allegedly may have given some innocent people cancer. Social justice, individual justice—I hoped we would be getting some of that here for Lee and for all the others.

After I entered, I joined Mike Miller and his entourage in the courtroom. It was the first federal courtroom I'd ever set foot in. Maybe not unexpectedly, given the scale of everything in the Burton building, it was a lot larger than I had expected. "They don't look this big on *Law and Order*," I said to myself.

I sat in the back, praying that no one could hear the *thump-thump-thump* of my heart. In the front of the courtroom, lawyers were scattered around two long tables. There were way too many laptops and yellow legal pads on both sides for me to count. The plaintiffs' lawyers were to my right, and the defense was to my left. I learned that the judge had allocated eleven hours in total to each side to make their case.

I recognized Judge Chhabria in the midst of this legal scrum. His youthful looks were belied by his complete command of the courtroom. In a Daubert hearing, as we've learned, the judge is the gatekeeper; there is no jury. Each side must convince the judge that their witnesses are genuine authorities on the topic about which they are testifying. I was sure Judge Chhabria was going to set a high bar in this case, which could turn on the often nuanced and rarely conclusive evidence of complex and technical scientific research. If you claimed to be an expert in Judge Chhabria's courtroom, you'd damn well better be prepared to demonstrate your expertise.

Chhabria earned my respect even before he questioned me. As he was speaking to the witness before me, I could tell from what he was asking that he had done his homework, had read all the briefs, and had extensively reviewed dozens of scientific articles. No one could fool this judge, I thought.

At that time, in March 2018, there were only 365 lawsuits combined in the MDL under the purview of Judge Chhabria. I say "only" because eventually the number of Roundup-related cases would reach the tens of thousands. The judge knew that number would grow, and because some of these cases would be heard in other jurisdictions in California, he had invited Alameda County Superior Court judge Ioana Petrou to join him for the Daubert hearing. Although none of the individual trials had even started yet, I learned that this would probably be one of the most important weeks of the entire litigation. After listening to every expert in this hearing, Judge Chhabria would decide who would be allowed to testify in future trials.

"You mean that we're sort of auditioning?" I had asked Mike Miller the evening before.

He laughed. "Chadi," he drawled, "y'all got the role as far as I'm concerned. I think the judge will agree, once he hears how much you know."

In other words, yes, it was an audition. A legal audition, in which the judge would determine whether I was a credible enough medical expert to be allowed to testify. Mike, Tim, and their associates would do their best to make me and the rest of our team sound like an assemblage of Nobel laureates, unassailable experts on the question of glyphosate and non-Hodgkin lymphoma. The Monsanto legal team, on the other hand, would, well, assail us. Basically, they would try to poke holes in the evidence in the hope that Judge Chhabria would throw the entire case out.

As I waited my turn, I listened intently to Dr. Beate Ritz, an epidemiologist from UCLA who was on our side. Dr. Ritz, whose testimony had begun the day before, was explaining the epidemiological literature and how to analyze data generated from such studies. Kathryn Forgie, a plaintiffs' attorney whom I would later work closely with, questioned Dr. Ritz in a methodical manner about the various studies that linked glyphosate to non-Hodgkin lymphoma. As Judge Chhabria listened carefully, periodically taking notes, Dr. Ritz explained why she believed that routine users, and not just farmers, have an increased risk of developing the disease. Every so often, the judge would take his dark-framed glasses off, place them on the table, and look deep into the courtroom, as if pondering what he had just heard. He would then turn and ask Dr. Ritz a question—and generally it was the right question, the logical question, the question that made it clear to her and everyone in the courtroom that he too had prepared for this hearing. He asked about regression models, confounders, and odds ratios. Very technical stuff. I was impressed.

But Ritz handled those questions well. She was convincing and articulate and had total command of the facts. Judge Chhabria expressed his concerns about the lack of adjustment to confounding factors in some of the studies, but she was able to explain to him confidently why lack of adjustment did not negate the other pieces of evidence that link the weed killer to lymphoma. "She's good," I thought to myself. "I need to be as cool and collected as she is."

Dr. Ritz went on to explain why the Agricultural Health Study (AHS)—the study that seemed to support Monsanto's side of this argument—was flawed. She reminded the judge, when prompted by Forgie, that there was animal data linking glyphosate to lymphoma.[2] And she offered compelling evidence that glyphosate could cause the DNA in cells to break, something that is often a precursor to cancer.

Judge Chhabria had referred to this Daubert hearing as "science week"—five days when the plaintiff and defense counsel had the opportunity to have their experts answer questions from the attorneys and the judge. Dr. Ritz and I were the two members of the plaintiff's expert team testifying on this day. The Miller Firm and the rest of the legal team had brought together a total of half a dozen medical experts representing the breadth of scientific and medical knowledge that would be needed to support the claims being made by Lee Johnson and the other plaintiffs.

Another of the expert witnesses testifying for the plaintiffs was Dr. Dennis Weisenburger, who at the time was the chair of the Department of Pathology at the City of Hope National Medical Center in Los Angeles County. During Weisenburger's testimony earlier in science week, Chhabria had asked him if he believed that real-life exposure to glyphosate can cause non-Hodgkin lymphoma in humans (as opposed to the deliberately high exposures used in animal studies). Weisenburger affirmed that the evidence was strong. He described studies showing DNA damage in people exposed to the weed killer, which proved biological plausibility. He voiced his criticism of the AHS. And he stated, "There is a body of evidence that is pretty compelling that glyphosate and the formulations are genotoxic in living cells."

My crafty old nemesis Kirby Griffis was the Monsanto lawyer who interrogated Weisenburger. Based on what I heard and read about their exchange, Griffis questioned Weisenburger on every study, every detail, every statistic. I had to hand it to him: he had done his homework. And he used that knowledge when cross-examining Dennis—apparently with some effect. According to journalist Carey Gillam, as Weisenburger left the witness stand, he was heard saying, "It's like going to hell and coming back." Yikes! Once again, I found myself wondering what I was getting myself into here.

Another key plaintiff witness during the Daubert hearing was Charles Jameson, who had served as a program leader for the National Toxicology Program at the NIH's National Institute of Environmental Health Sciences for twelve years. More importantly, he was a member of the working group for the IARC that found glyphosate to be a probable human carcinogen. His involvement in this litigation was crucial since he had reviewed the evidence that led to the IARC determination. Jameson explained the famous "mouse study" to Judge Chhabria. Monsanto had tried to keep much of the data from this study out of court, but the judge ruled that the study data was relevant and would be allowed into evidence. Importantly, Jameson summarized the animal studies and commented that tumors were seen repeatedly in specific sites in exposed animals, including liver tumors and malignant lymphoma. Judge Chhabria listened

carefully and then questioned Jameson about the IARC classification. Could it be interpreted as suggesting that there was
"limited" evidence of carcinogenicity in humans? Jameson
responded that while some scientists at IARC thought the evidence was strong, others disagreed. This, he said, was not uncommon in his profession. "If there are three epidemiologists
in a room and you ask them their opinions, you'll get four opinions," he said.

I chuckled when I heard that. Sounded to me like every other
medical specialty as well. And it was a reminder that nothing was
ever 100 percent certain in scientific research. The only absolute in
my mind was that Monsanto was wrong in the way they'd handled
this. Why couldn't they acknowledge the IARC findings (even if
they disagreed with them), and warn patients that some might be
affected? I don't know.

Jameson was cross-examined by Monsanto's counsel Joel
Hollingsworth, whose continual probing about the epidemiologist's responses to questions in prior depositions did not please
Judge Chhabria. "Why don't you ask him about his opinion *now*,"
he barked at Hollingsworth. "That's normally how we do it." But
Hollingsworth continued to dissect statements that Jameson
had made in the past. The judge again instructed the attorney
to ask Jameson about his present opinions, and if what he said
contradicted something he had said earlier, Hollingsworth could
then explore the contradiction. Judge Chhabria also criticized
Hollingsworth for talking over Jameson as the latter tried to answer questions. Moreover, the judge showed concern over the
possibility that Monsanto might be taking expert statements out
of context. That concern was underscored when, in a particularly
stern move that left the plaintiff attorneys giddy with delight,
Judge Chhabria ordered Monsanto's attorney to read aloud into
the record two pages of testimony from a deposition that supported Jameson's analysis, before he would allow Hollingsworth

to introduce a separate example from a deposition that undercut Jameson's expertise.

Christopher Portier was another one of the plaintiff experts. At the time of the litigation, he was retired and living in Switzerland. Portier had been present (but was a non-voting participant) at the IARC meeting in March 2015 when glyphosate was discussed. Before retiring, he had led the National Center for Environmental Health/Agency for Toxic Substances and Disease Registry at the Centers for Disease Control and Prevention (CDC). Portier was questioned by Robin Greenwald, another one of the lead plaintiff attorneys whom I closely worked with afterward. Robin has a quiet and empathetic demeanor that appeals to judges and juries alike. You could sense that she deeply cared about the patients who were allegedly affected by Monsanto's product.

Portier addressed the controversial role of the EPA in this saga. He essentially advised Judge Chhabria that the agency had not followed their own methodologies and guidelines in assessing the carcinogenicity of glyphosate. He reviewed the animal studies again and explained why he believed that there exists a causal relationship between glyphosate and non-Hodgkin lymphoma. It was becoming clearer that determining the cause-and-effect relationship between glyphosate and non-Hodgkin lymphoma would require examining the totality of available evidence, not just one or two studies. I hoped I'd get the opportunity to reaffirm that point.

Dr. Ritz's testimony concluded, and she stepped down from the witness stand. Now it was my turn.

I walked as confidently as I could—standing straight and tall—towards the witness stand, as the eyes of about twenty attorneys followed me. It was unnerving. It was funny, I thought as I took my seat. In all the legal thrillers I'd watched, witnesses were always composed, unfazed. Fake TV again, I thought.

After being sworn in, I was first examined by Mike Miller. We discussed the epidemiological literature again, but from a clinical standpoint—meaning how such data impacts patient care and counseling. Very few clinicians are trained as epidemiologists, yet we need to understand such literature so that we can help patients. Miller asked me about the latency period, which, as we've seen, is the time it takes from initial exposure to a hazardous compound until the development of cancer. I provided a few examples of how latency varies in non-Hodgkin lymphoma. I presented studies I'd found in which lymphoma had developed just months after patients were exposed to a hazardous agent; in other studies, and with other agents, it took a few years. Basically, I made the case that latency periods vary in duration. As I was presenting some studies that didn't involve glyphosate, for illustrative purposes, I got the sense that Judge Chhabria wasn't pleased. Whereas his attention had seemed riveted on Dr. Ritz, I noticed he was reading some papers as I spoke. Was he listening? I should have taken that as a signal, but I felt that I was trying to make an important point, which is that some cancers have a short latency period, while others have longer ones. Some smokers, to take one example, might develop lung cancer shortly after starting smoking, while for others it might take years or decades—or they might not develop it at all. While we clinicians try to take the epidemiological literature into account, the patients that we treat are not statistics, they're individuals. I was trying to convey that we need to look at every case separately and not ignore cases where the latency period might be short.

The judge did not appear impressed by my argument. "We are discussing pesticides and glyphosate," he said curtly. "I'd like to hear evidence on latency for these." There was little available— a fact that I sensed he knew—and when I said as much, he responded with a shrug, as if we'd wasted enough time on me.

Judge Petrou, who was seated on the bench alongside Chhabria, came across as more pleasant despite her not speaking much, and I sensed that she connected more with what I had to say. But of course, this was not her courtroom; she had been invited only to observe.

It was bad enough that I'd seemed to have blown my appearance with Judge Chhabria. Now I had to deal with Griffis again. He employed the same strategies he'd used during the depositions. First, he focused on my MBA. To hear him explain it, I'd practically committed a crime by going to business school, and no longer cared about patients. He attempted to paint a picture of me as a "suit" who knew nothing about clinical medicine— as if the two decades I'd spent working with patients with non-Hodgkin lymphoma were nullified because I had added the letters "MBA" after "MD."

Then Griffis pivoted to the latency period issue that had already been discussed when Miller examined me. Griffis claimed it should exceed twenty to twenty-five years, an argument that I felt was illogical. Again, to use tobacco as an analogy, you can't tell someone that latency is always five years, or ten, or twenty. You can't say that if someone smokes less than one year, they'll be fine, but if they smoke for more than one year, they'll get cancer. There are no absolutes for a clinician like me. Every patient is a case study of one. Lee Johnson had worked with glyphosate five days a week for two years; then he developed his lymphoma. In his case, two years of consistent use certainly seemed a reasonable latency period. And that's the kind of judgment a clinician needs to make.

Griffis then went on to sound a familiar refrain from the Monsanto playbook. He dismissed the epidemiological literature I relied on, saying that it was not complete and claiming there were too many confounding factors to make any kind of definitive interpretation. The glaring exception was the AHS. This, if

Griffis and his colleagues were to be believed, had no flaws. It was the perfect study. Or so they tried to make Judge Chhabria believe. On the flip side, Griffis criticized *my* methodology and argued that I was relying only on the IARC report and had not reviewed all of the epidemiological studies myself.

That was an out-and-out falsehood, and I tried to contain my frustration as he pursued that argument. I knew how many hours, days, nights, weeks, and months I'd spent studying all the literature about glyphosate and lymphoma. My own family could attest to how much I'd worked on this. And now Monsanto was claiming to a federal judge that I relied only on one report and one study? Yet, to my dismay, I felt that their argument resonated well with Judge Chhabria. Griffis showed the judge several transcripts from my prior depositions where I had clearly stated how important the IARC report was and that I had relied "heavily" on it. That was taken out of context in my opinion, but it didn't matter. They had it on record.

Judge Chhabria nodded and took some notes as Griffis continued his arguments. As I saw the judge's reaction, my heart sank. Lesson learned: I needed to be much more careful in articulating my views in the future. Every single word matters.

After four hours of being on the stand, I was finally allowed to get down. The Uber driver pulling up to the entrance of the Burton Building to take me to the airport was a welcome sight. I arrived in Chicago close to midnight. Everyone was asleep when I entered my home. I glanced at my kids, who were each sleeping sideways in their own beds, kissed them on the forehead, and crashed in my bed near my wife, who was also deeply asleep.

A few weeks later, I learned that Judge Chhabria would allow my testimony on a case-specific basis, but not in the general causation. That meant that I could testify and offer my opinion in individual patient cases (such as Lee Johnson's) and explain why cancer might have developed in that particular patient's case, but I could not serve as an expert witness on the general causality of glyphosate and cancer.

The ruling hurled me into a black mood. But the self-pity didn't last too long. I reminded myself that I still had a part to play in this courtroom drama. I had to do better when I got my next opportunity to help get the justice that Lee Johnson, and the others, deserved. That was what mattered the most.

With that in mind, I was more determined than ever to continue to be a part of the legal fight against Monsanto.

8

The Johnson Trial Begins

As the plane's tires hit the tarmac with a loud bang, I held on to the seat in front of me for dear life. There were mutterings among the passengers as we rolled toward the gate and the unfazed-sounding voice of the flight attendant bid us welcome to San Francisco, where I would be testifying in the first Roundup trial against Monsanto. I wondered if the trial itself would go more smoothly than the touchdown.

It was now July 2018, four months following the Daubert hearing, and the Boeing 757 United flight 222 that I was on landed on a warm day at the San Francisco airport. Tim Litzenburg, wearing a T-shirt and blue jeans, was waiting for me outside of the baggage claim area.

"Where's your luggage?" he asked.

I held up my oversized briefcase.

"That's all you need?"

"For a one- or two-day trip? I'm surprised you don't travel like this," I kidded him. "All you seem to need is a couple of T-shirts."

He laughed. "Well, I won't be wearing a T-shirt when we appear in court, and as a lawyer, I would advise you to wear your best suit."

I smiled. "It's right in here," I said, and gestured toward my briefcase. "And I packed my power tie."

When I got into Tim's rental car, empty coffee cups, papers, and travel documents were strewn all over the place. "Even this car you manage to mess up," I teased.

"I do my best thinking when I'm surrounded by clutter," he said with a grin. I shoved some papers out of the way before I sat down. I didn't mind, though—it was all part of Tim's charm.

That day, Tim informed me as he turned the car onto US-101, we were driving to the "War Room"—the nickname the Miller team and other plaintiff attorneys had given to the office they were borrowing from a local law firm. Tim updated me on what had been happening in the trial, which had begun a few days earlier. "Things are looking good," he said. "The jury is learning about the science, and they've seen some of Monsanto's internal correspondences."

I was among the last witnesses that the plaintiff lawyers were calling to testify. Prior to my appearance, the jury had heard opening statements from the lawyers and testimony from several other medical experts.

As Tim drove, my thoughts were racing. Only hours separated me from facing the judge and the jury. As it was one of the largest lawsuits ever brought against a corporation, and because most Americans had either used or heard of Roundup, a ubiquitous product in Home Depot and other retailers from coast to coast, there was intense media interest in this trial.

And then there was the particular nature of the trial. This wasn't a cohort of nameless individuals grouped into a class action suit. This was an ordinary working guy, Lee Johnson, standing up to a corporation whose earnings had been over $14 billion the year before the trial.[1] It was like David and Goliath, and much like the biblical David, Lee was fighting for his life against this giant.

Many of the reporters covering the trial had immediately recognized the power of this story line, and I suspected that if we could have surveyed people about this ongoing litigation, the vast majority of Americans would have been on Lee's side. *Our* side, I added to myself. I was ready to do my part to help Lee get justice.

In the War Room, more than half a dozen people—lawyers and paralegals from the four different firms that had come together to represent Lee Johnson and plaintiffs in other lawsuits—were arrayed in front of their laptops around an oval dark-colored oak table. Along one wall, shelves groaned with files. A large picture window on the other side of the room looked out at the city. The building we were in was on Van Ness Avenue, near the Civic Center area, the epicenter of power in San Francisco and just a few blocks from the Phillip Burton Federal Building, where I'd testified at the Daubert hearing.

Tim introduced me to two of the other key players in this trial: David Dickens and Brent Wisner, the co-counsel who were trying the case on behalf of Lee Johnson. Dickens was another attorney in the Miller Firm, and Wisner was a Los Angeles–based attorney with Baum Hedlund.

One person I didn't see there was Mike Miller. When I asked where he was, they told me that Mike had been in a boating accident a couple of weeks earlier in which he'd suffered a punctured lung, precluding him from traveling and trying the case. It was a freak accident, and while Mike was extremely frustrated that he couldn't be present, he was lucky to be alive. I thought it was rather unfortunate that his accident had happened not long before this first-of-its-kind trial. I got a bit more nervous when Tim mentioned to me later that he'd had a seizure just a few weeks before. I was alarmed to hear this. "Are you okay?" I asked. "I trust your doctor cleared you to travel?"

"I'm fine," he said. "But they still don't know what caused it."

The head of the law firm gets into a boating accident just a few weeks before the trial—and then, out of nowhere, one of his associates has a seizure? This really *was* beginning to sound like a John Grisham novel. I'm not prone to conspiracy theories, however, and soon I dismissed the idea that this was anything more than a bizarre coincidence.

"Tim," I said, joking with him about something I normally wouldn't joke about, "I'm going to keep my eyes out for snipers as we walk around the city."

Both of us chuckled. Of course, there was no sinister plot. (However, I must admit that I never told my wife any of this.)

That afternoon, I also learned that Monsanto had hired a prominent litigator named George Lombardi as their lead defense counsel for this case. I had expected I might again be matched against my familiar adversary Kirby Griffis, whom I'd seen three or four times before, but no, it was Lombardi who would be cross-examining me. I immediately started internet sleuthing and read as much about him as I could find. I learned that *American Lawyer* magazine had named him "Litigator of the Year" in 2014, which was unnerving to read the day before I faced off with him. He had earned that distinction for what was described as "his staggering win in *Monsanto v. Dupont,* which resulted in a $1 billion jury verdict."[2] I also learned that Lombardi had "represented tobacco companies in their litigation cases."[3] Clearly, Monsanto did not care much about the optics. They believed that Lombardi as their lead counsel provided them with the best opportunity to win, and they went for it. Brent and David told me that Lombardi connected very well with juries and was cool under pressure.

I guess you need to be that good to be named Litigator of the Year. I guess you need to be that good to represent Monsanto in a case like this.

* * *

The trial started on July 9, 2018, amid heavy media coverage. "Does Roundup Cause Cancer?" screamed a headline in the *San Francisco Chronicle* that morning. Judge Suzanne Ramos Bolanos was presiding, and I learned from the other attorneys that the assessment Tim had given me the previous day of how the case was going so far was a bit optimistic. Judge Bolanos had denied several motions by the plaintiff counsel that they felt were critical for the jury's understanding of the full picture.

Opening statements in any trial are about drawing a road map for the jury of what they will see and hear. The lawyers on both sides usually explain to the jury what the trial is about and what they will try to prove through their witnesses and evidence. Lawyers' opinions are not considered evidence, and judges explain this explicitly to the jury. Attorneys can say what they wish, within certain limitations, during opening statements, but in the final analysis, these statements will need to be supported by evidence. The defense, on the other hand, doesn't need to prove anything. They simply need to poke holes in the plaintiff's arguments. However, the plaintiff lawyers must prove their case to the jury by providing factual and legally admissible evidence to support their claims. Customarily, the plaintiff starts by presenting their case, and the defense has an opportunity to cross-examine the plaintiff witnesses. This is followed by the defense presenting their case and calling their witnesses; the plaintiff attorneys then have an opportunity to cross-examine the defense witnesses.

Brent Wisner made the opening statement on behalf of Lee Johnson, who was in the courtroom that day, and did so with a powerful and eloquent argument about what was at stake (see figure 14). "This case really is about choice," Brent said. "It's about the right of every single person in this room to make a choice about what chemicals they expose themselves, their family, or their children to. And, sure, some people are willing to take

Figure 14: (*Left*) Brent Wisner, of Baum Hedlund, during the Johnson trial. (*Right*) David Dickens, of the Miller Firm, during the Johnson trial; to his right is the plaintiff, Lee Johnson. Wisner and Dickens were the co-lead counsel for the plaintiff in the first-ever jury trial against Monsanto. Source: Reuters/Alamy Stock Photo

greater risks. After all, people still smoke, and we all know smoking causes cancer. But at the end of the day, it's our choice. And really nobody has a right to take that choice away from us simply because they would deprive us of information that we need to make that choice."

He made sure that the jury understood that Ranger Pro, the product that Lee Johnson had used on the job, contains glyphosate, the same active ingredient as Roundup. He spent some time explaining how glyphosate is mixed with surfactants to make the final product. In his witty and eloquent manner, he pointed out that while Monsanto always asserts that glyphosate doesn't cause cancer, in its regulatory filings it never claims that *Roundup* doesn't cause cancer. It's a subtle but significant distinction.

He also summarized animal studies that were done on mice (but not by Monsanto) and showed the jury that three of these studies demonstrated an increased incidence of kidney tumors, while four showed malignant lymphomas. He spent a few minutes teaching the jury how glyphosate could cause cancer, and then summarized most of the epidemiological studies I've described in previous chapters. He did not shy away from criticizing the Agricultural Health Study (AHS), which Monsanto relied so heavily on.

Wisner knew that Monsanto would also rely on the fact that the EPA did not classify glyphosate as carcinogenic, so he addressed that point without overtly criticizing the agency. "The EPA can only see what's given to them," he said, referring to the study by British researcher James Parry, who had been hired as a consultant by Monsanto to assess the carcinogenicity of glyphosate and was subsequently dropped when he recommended more studies (which, as we've seen, were never conducted). "We know Dr. Parry's report was never given to the EPA." Wisner also reminded the jury about the realities of life in Washington, DC: "Like any political agency, the EPA ultimately is subject to political shifts. I'm not suggesting one political side or the other is at play here. Okay?" He paused for a few seconds and looked at the jury, making sure they were following him. "But I will say that it's relevant, right? Because sometimes political decisions can trump scientific decisions. That's just something we should be thoughtful of."

Wisner's core argument centered around the fact that while Monsanto's scientists and executives knew of the association between non-Hodgkin lymphoma and Roundup, they not only ignored the facts but also implemented strategies to hide and distort these facts.

Because I am a physician trained to respect the scientific process and the literature that grew out of that process, this argument had certainly resonated with me from the get-go. I was disappointed by Monsanto's efforts to downplay any scientific evidence that showed a link between glyphosate and non-Hodgkin lymphoma. Not acknowledging the science had another effect: it robbed Lee of the right to make an informed choice. Had Lee known the potential carcinogenicity of Roundup, he might not have used it. Had his employer known, they might have implemented additional measures to mitigate exposure. Knowledge, as they say, is power. Similarly, Lee and others could have still chosen to spray and be exposed despite knowing that a risk existed. The bottom line: individuals should decide what their tolerance for risk is, but they can't unless they are fully informed. I have had patients with cancer decline my recommended treatment because of side effects they were unwilling to endure. I vividly recall a patient who refused chemotherapy because the treatment would have led to severe hair loss. While I might not have agreed with that decision, I respect that my patient had made an informed choice. We all want the liberty to make our own decisions. That's what America is all about, right?

Wisner again reminded the jury of the Parry report and how Monsanto admitted that it had no record of having submitted that report to the EPA. Wisner went on to share with the jury that they would see during the trial various email communications showing how Monsanto had engaged in ghostwriting scientific articles, especially after the IARC report came out, and how they had planned and orchestrated a media campaign to proclaim Roundup safe after the final IARC decision was made public in March 2015.

Monsanto's lead attorney, Lombardi, was ready with his response in Monsanto's opening statement. (See figure 15.) He reminded the jury that the burden of proof rested on the shoulders of the plaintiffs. He effectively made the point that the plaintiffs would fail to present convincing scientific evidence that glyphosate causes cancer—because, he claimed, there was none. He preemptively poked holes in the epidemiological studies, saying that these studies did not adequately account for exposures to other possible carcinogens. He emphasized that the EPA, as well as other regulatory agencies across the world, had deemed glyphosate safe, noting that the EPA had considered sixty-three epidemiological studies involving the chemical (although, I knew, some of them were not related to lymphomas). He also downplayed Wisner's argument about the politics of the EPA. "Counsel made some reference to the EPA being subject

Figure 15: George Lombardi, lead counsel for Monsanto, in the Johnson trial, July 2018. Source: Reuters/Alamy Stock Photo

to politics," Lombardi said. "And clearly the top of the EPA is subject to politics, but what's striking about the history here is that the EPA has come to the same conclusion through different administrations, through different generations of EPA scientists, through the accumulation of testing. The science has been consistent that glyphosate is not carcinogenic."

Lombardi claimed that Monsanto did indeed conduct all the animal and cellular studies that were required and suggested that the EPA was more inclusive in the studies they reviewed than the IARC was. Monsanto had maintained all along that there were more than 140 studies that supported the safety of their compound, and Lombardi pointed a finger at Wisner for cherry-picking the studies that suited him. (Considering how Monsanto had clung to the AHS while ignoring so many other studies, I thought this was like the pot calling the kettle black.) He asserted that there had been no ghostwriting, and that Monsanto did declare when they had any input in the published articles. The last slide in his PowerPoint presentation showed pictures of

Johnson's physicians. None of the doctors who cared for Lee, he declared, had ever suggested that Roundup had a role in causing Lee's lymphoma.

This was a point that Monsanto emphasized frequently in this trial. But it's easily explained. Lee's doctors were more focused on the treatment of his cancer. They may not have been aware of the link between Roundup and lymphoma. Heck, until I got involved in this case and reviewed the epidemiological literature, I didn't know that, either.

All of this had played out a few days before my arrival in San Francisco. As I read the news reports and got firsthand accounts from the attorneys in the War Room, I could see this was going to be a tough fight. But I also had a question: would any Monsanto employees be testifying in the courtroom? To my surprise, the answer was no. My first reaction was surprise: were *none* of the company's 20,000-plus employees willing to come on the stand and defend these accusations? It turned out that there was a legal reason involved: because the trial was in California and Monsanto's headquarters were in St. Louis, Missouri, the plaintiff lawyers could not demand the live presence of any Monsanto employee. Still, my initial reaction suggested that, even if there were legalities involved, the optics were bad, and that perhaps members of the jury might have some of the same thoughts. If I had been a jury member, might the fact that I wasn't seeing at least a few employees of Monsanto defend the company's reputation and integrity in court suggest that the company lacked either? I wouldn't know, but these thoughts did cross my mind.

This was not lost on Wisner, who offered some memorable tongue-in-cheek comments in his opening statement about the absence of anyone from Monsanto in the trial. "There is a big limitation in this case," he said to the jury. "And that is the fact that there are ten different individuals, former or current employees of Monsanto, who are going to testify, and none of them are going to physically appear live. I can't make them." He went on to explain to the jury that they would be watching video depositions of these employees—and hinted that these would be neither useful nor interesting. "You're actually going to see a lot of videos," he said with a mischievous grin. "I mean, a *staggering* amount of video. I apologize for that. It is so unbelievably boring and dry; it makes me want to cry. But it just is what it is."

As part of my preparation for the trial, I had read all these Monsanto employees' depositions myself. While it was some-times a challenge to follow all the back-and-forth, it gave me a better understanding of many of the talking points that Mon-santo was likely to make—and of the questions I might be asked when it came time for me to take the stand. And this time I wanted to be as thoroughly prepared as possible. While I knew that Mike, Tim, David, and the rest of the legal team had seen my Daubert hearing performance as a success, I was still smart-ing over the fact that all the research I'd done had somehow been undervalued that day. I wasn't going to let any Monsanto lawyers get the better of me again.

One of the salient issues that were raised in the first few days of the trial was the difference between the herbicide glyphosate and the commercially sold product Roundup. In Roundup, glypho-sate is mixed with substances, called surfactants, that are de-signed to enhance the weed's absorption of the herbicide. These surfactants also increase glyphosate absorption when Roundup comes in contact with human skin. Wisner explained to the jury that the surfactant known as polyoxyethyleneamine (POEA), used in the US formulations of Roundup, is in fact banned in Eu-rope because of concerns over its toxicity.[4] As I've noted earlier, Monsanto had always argued that glyphosate does not cause cancer, citing some of their own earlier studies—but had admit-ted many times that they had never conducted carcinogenicity studies on the formulated product, meaning glyphosate plus the particular surfactants in Roundup. To me, this was an important point that should not be overlooked.

In the days that preceded my testimony, the plaintiff attorneys had explained to the jury the importance of looking at the totality of the evidence. They were urged to take into consideration the evidence from animal studies, the information about how glyphosate could damage cells by disrupting their DNA, and the human epidemiological research. My role would not be to get into the specifics of any of these studies; rather, I was to explain to the jury the practical implications of this research, showing how clinicians like myself use such evidence to help counsel patients who are facing this disease—patients like Lee Johnson.

The day before my appearance, Christopher Portier, the former CDC official who had been present at the IARC meeting in March 2015 when glyphosate was discussed, took the stand and described in detail the various animal and toxicology studies. Portier explained to the jury that there were several working groups within IARC, one of which focused on animal studies and another on mechanistic data; the data that proposed how glyphosate might cause cellular and DNA damage.[5] Portier explained to the jury the different animal studies that showed the potential carcinogenicity of glyphosate and acknowledged that none of these studies had been done on Roundup. One 1983 study, for example, showed that exposure to glyphosate produced increased rates of kidney tumors and lymphomas of the spleen. In four additional studies conducted between 1990 and 2010, there was increased incidence of lymphoma in each, with three of them also duplicating earlier findings of increased risk of kidney tumors. Another important point was that the jury learned of a study conducted in 2010 in which mice exposed to glyphosate demonstrated a significant increase of tumor formation.[6] About 40 percent of the treated animals in this study by J. George and colleagues developed tumors; none of the mice in the control group did. The EPA had dismissed that study on the grounds that fifty mice should have been studied, not just twenty. Interesting![7] Portier also described proposed mechanisms by which glyphosate could cause cancer. One of them involves genotoxicity, or damage to a cell's DNA. Another involves oxidative stress, a breakdown in the cellular repair process. It's noteworthy that Monsanto did agree that these two mechanisms could potentially cause cancer, but of course they denied that their product could exhibit these two mechanisms. Portier reviewed several dozen genotoxicity studies and told the jury that real-world exposure can cause DNA damage.

Portier was also not shy about disagreeing with the EPA's classification of glyphosate and with how the European Food Safety Authority (EFSA) had evaluated the carcinogenicity of glyphosate. In this excerpt of his exchange with Kirby Griffis, he refers to bioassays, which are defined simply as a measurement of the concentration or potency of a substance in terms of its effect on living cells or tissues. Any substance classified by the EFSA as 1B is presumed to have carcinogenic potential for humans.

Kirby Griffis: *You testified earlier today that EPA was so amazingly wrong in its evaluation of glyphosate and the conclusions it came to.*

Christopher Portier: *The eventual conclusion was that there's no evidence supporting carcinogenicity or something along those lines.*

KG: *Yes.*

CP: *I find that conclusion astonishing.*

KG: *Do you also find EFSA's conclusion astonishing and so amazingly wrong?*

CP: *Yes, I do. By their own guidelines, which clearly state, for example, that if you see two positive animal bioassays it should be classified 1B. That's all that's required for their 1B classification, by their own guidelines. And I've demonstrated over and over*

again that there are much more than two positive findings in these
data. Hence, they have not followed their guidelines, and so I do
disagree with them.
KG: *They do know what their guidelines are; right?*
Brent Wisner: *Objection. Speculation.*
CP: *I know what their guidelines are. So do they.*
Court: *Overruled.*

When Wisner questioned Portier, they reviewed the guidelines by which EFSA and the EPA assess carcinogenicity and how in the case of glyphosate neither followed their own guidelines. When Wisner asked Portier why he believed that the EPA classification was incorrect, Portier answered that to him the evidence that glyphosate is carcinogenic is "so overwhelming." By contrast, the EPA put glyphosate into the category of substances "not likely to be carcinogenic to humans." Portier said that in his opinion, putting a substance into this category means that "you have evidence where virtually everything's negative. There's just nothing there that would support a carcinogenic finding, and you have a lot of evidence. And so, you'd say, you know, I'm pretty comfortable with saying this is not likely to be carcinogenic to humans. That's how—that's how you would put it into that category."

In fact, in 2016, Portier and several colleagues published a commentary to EFSA explaining their rationale for disagreeing with EFSA's carcinogenicity assessment of glyphosate. In the conclusion of that letter, the authors state, "The most appropriate and scientifically based evaluation of the cancers reported in humans and laboratory animals as well as supportive mechanistic data is that glyphosate is a probable human carcinogen. On the basis of this conclusion and in the absence of evidence to the contrary, it is reasonable to conclude that glyphosate formulations should also be considered likely human carcinogens."[8]

Although it included discussions of some very technical aspects of the evidence and the regulatory process, Portier's testimony seemed clear and unambiguous, at least to me. I hoped the jury had grasped it. The lawyers knew that you could lose a jury if you get too far into the weeds—and I don't mean the ones that Roundup was designed to eradicate. Explaining complex science in an easy-to-understand language is essential to winning over the jury and the trial. When it was my turn to testify, I planned to try to explain the facts and the science to the jury in the same way I would to my patients and their families.

Prior to my testimony, the jury had already heard about the alleged "ghostwriting" that Monsanto had engaged in, aimed at countering IARC's conclusions and the scientific facts. A number of Monsanto's internal communications that were made available through discovery were shared with the jury. Journalist Cary Gillam and her collaborators Nathan Donley and Sheldon Krimsky subsequently chronicled the details of this practice by Monsanto. The practice of ghostwriting, Krimsky wrote in an August 2018 essay in *Environmental Health News*, which has been "largely disavowed by respected journals as a form of plagiarism, appears as a normal business practice for Monsanto."[9]

A flagrant example of this was found in the scientific journal *Critical Reviews in Toxicology*, which in September 2016 published a series of papers assessing the carcinogenicity of glyphosate.[10] These papers were widely cited by media outlets because they opposed the IARC findings and proclaimed that glyphosate was safe and does not cause cancer. But the editors and reporters at these news outlets never knew that Monsanto was behind the writing of these scientific papers that were favorable to Monsanto's views. Of course, none of Monsanto's alleged involvement in these papers was properly declared, in direct contravention of the publisher's conflict-of-interest policies.

The most influential of these papers was the one written by Gary M. Williams and colleagues. Here, sixteen scientists declared that glyphosate was not carcinogenic, and that this conclusion was independent of Monsanto's influence. "Neither any Monsanto company employees nor any attorneys reviewed any of the expert panel's manuscripts prior to submission to the journal," they wrote at the end of their paper.[11] A powerful assertion by the authors—except it was simply incorrect. Internal email communications made publicly available show how a Monsanto scientific team was heavily involved in analyzing the topics, drafting, editing, and approving the conclusions. (See figures 16 and 17.) Even when one of Monsanto's scientists, John Acquavella (who added the comments marked as "JA" in the document reproduced here), suggested toning down the inflammatory statements against IARC in one of the articles, he was overruled by William Heydens (in the comments marked "wh"), who clearly urged the paper's authors to "ignore John's comment."

Heydens was questioned about this by Mike Miller in a January 2017 deposition, which was one of those shown to the jury. Under oath, he stated that the manuscripts had been sent to him, but that he'd only read parts of them. Strangely enough, he declared that he was unable to recall whether he had made any of the twenty-eight edits that Miller and the other plaintiff attorneys counted on the internal documents.

When emails related to these manuscripts became public and more people were made aware of the alleged ghostwriting, the Center for Biological Diversity sent a letter to the journal and the journal's publisher, Taylor & Francis, explaining their concerns, highlighting the clear ethical misconduct by Monsanto and its employees, and demanding retraction of those articles.[12] Charles Whalley, the managing editor for Taylor & Francis's medicine and health journals, proposed retracting the three

---

Message

From:	HEYDENS, WILLIAM F [AG/1000] [/o=Monsanto/ou=NA-1000-01/cn=Recipients/cn=230737]
on behalf of	HEYDENS, WILLIAM F [AG/1000]
Sent:	1/13/2016 11:26:42 PM
To:	'Ashley Roberts Intertek'
Subject:	RE: Summary report
Attachments:	Combined Manuscript DRAFT JAN 11 2016 (3) wfh review.docx

Hi Ashley,

Here are my suggested edits to the Draft Combined Manuscript. Most of my edits were made in Section 3.1 (Exposures to Glyphosate), as it read like a repeat of the entire Results section from Keith's Exposure paper/chapter, including table/graph replication as also noted in John Acquavella's email.

One thing I noted right off the bat was the order of tackling the 4 areas – 1) Exposure, 2) Animal, 3) Genetox/MOA, 4) Epid. This is different than IARC and different than what I thought we discussed, but I'm not opposed if you/others think this is the best overall flow.

Also, just FYI, it appears that your writer did not have the latest version of Keith's paper, as I found some differences which I confirmed with Marian Bleeke – I think I caught all the differences and made the changes in the Combined Manuscript as part of my editing. And I am going to thoroughly read the latest version of Keith's paper tomorrow; but now I am not inclined to suggest substantial re-writing (adding of text) because I don't want to slow progress down any more than necessary (my management would love to get all this off to CRT/Roger by very early February).

As an aside, I was struck by how similar the criticism of IARC in today's EFSA response to Portier was to points made by the expert panel – I think they are very closely meshed and complement each other nicely.

Anyway, let me know if you have any questions or concerns regarding my suggested edits.

Thanks much,

Bill

Figure 16: Internal email communication (made publicly available) shows that Monsanto was involved in significant editing. Note the dates of the emails, in January 2016. (*continues*)

articles that had claimed glyphosate was safe. Roger McClellan, the editor in chief of *Critical Reviews in Toxicology*, pushed back in an email that was made public, asserting that the papers were scientifically sound and had been produced without external

From: Ashley Roberts Intertek ███████████████
Sent: Monday, January 11, 2016 12:35 PM
To: HEYDENS, WILLIAM F [AG/1000]
Subject: Summary report

Hi Bill,

Please find attached the summary report for your review.

If okay we will add in the references and have wordpro format properly.

Thanks

Ashley

Ashley Roberts, Ph.D.
Senior Vice President
Food & Nutrition Group
Intertek Scientific & Regulatory Consultancy
███████████
███████████ @ ███████
███████████████

From: HEYDENS, WILLIAM F [AG/1000] ████████████████
Sent: January-11-16 8:35 AM
To: Ashley Roberts Intertek
Subject: FW: Publication Plans

MONGLY00998683

Figure 16: Continued

influence. A retraction, he argued, would tarnish the reputation of the authors, the journal, the publisher, key employees, and of course himself. In a later email about six weeks prior to the start of the Johnson trial, McClellan admitted that these papers were of "vested interest" to Monsanto. Nevertheless, he argued that

Ashley, this is the quick response I sent back to Larry Saturday. So far, I have not heard back from him.

From: HEYDENS, WILLIAM F [AG/1000]
Sent: Saturday, January 09, 2016 1:07 PM
To: Larry Kier
Subject: Re: Publication Plans

Hi Larry,

The current concept is 6 papers back to back in a single issue, possibly a stand alone supplement:
Introduction
Overall comprehensive summary
Epidemiology
Animal bioassays
Genotoxicity
Exposure

The order could change.

This is what Ashley and I thought would work best, and Roger McClellan seemed to agree in a preliminary conversation.

Do you see a problem with this?

Thanks

Bill

-------- Original Message --------
Subject: Publication Plans
From: Larry Kier ▓▓▓▓▓ o
Date: Jan 9, 2016, 8:15 AM
To: "HEYDENS, WILLIAM F [AG/1000]" ▓▓▓▓▓▓▓▓▓▓▓▓▓

Bill,

One of our panelists inquired about the publication plans because they were told at the August meeting at Intertek that
the panel report would be published as one paper with different sections and It now appears there will be separate
papers.

Could you or Ashley please clarify the current publication plans?

Thanks.

Figure 16: Continued

"the five papers are scientifically sound," and he asserted that "it would be a breach of scientific ethics and my own standards of scientific integrity to agree to retraction of any or all the glyphosate papers."[13]

Whalley, the group editor, disagreed, and said that the authors were in violation of publishing ethics. Moreover, he added, retracting the papers would show that the journal's editorial policies, with their insistence on declaring potential conflicts of interest, were working.

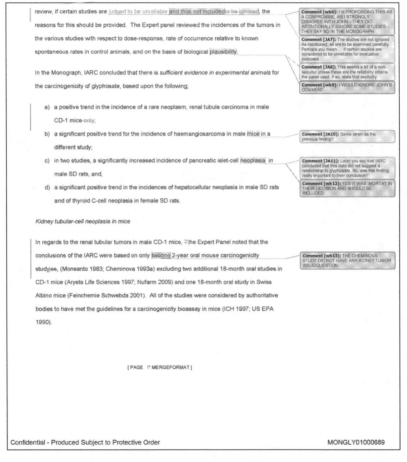

MONGLY01000689

Figure 17: Comments from Monsanto scientist John Acquavella (with the initials "JA") on a draft glyphosate paper, with comments from Monsanto regulatory science chief William F. Heydens (with the initials "wh") asking the paper's authors to ignore one of Acquavella's comments on the IARC and keep a phrase Acquavella had suggested deleting. Heydens controlled the tone of the paper but was not listed as a co-author. (*continues*)

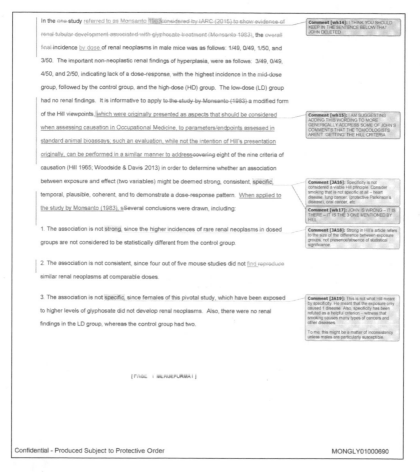

In the one study referred to as Monsanto 1983 considered by IARC (2015) to show evidence of renal tubular development associated with glyphosate treatment (Monsanto 1983), the overall final incidence by dose of renal neoplasms in male mice was as follows: 1/49, 0/49, 1/50, and 3/50. The important non-neoplastic renal findings of hyperplasia, were as follows: 3/49, 0/49, 4/50, and 2/50, indicating lack of a dose-response, with the highest incidence in the mid-dose group, followed by the control group, and the high-dose (HD) group. The low-dose (LD) group had no renal findings. It is informative to apply to the study by Monsanto (1983) a modified form of the Hill viewpoints, which were originally presented as aspects that should be considered when assessing causation in Occupational Medicine, to parameters/endpoints assessed in standard animal bioassays; such an evaluation, while not the intention of Hill's presentation originally, can be performed in a similar manner to address covering eight of the nine criteria of causation (Hill 1965; Woodside & Davis 2013) in order to determine whether an association between exposure and effect (two variables) might be deemed strong, consistent, specific, temporal, plausible, coherent, and to demonstrate a dose-response pattern. When applied to the study by Monsanto (1983), sSeveral conclusions were drawn, including:

1. The association is not strong, since the higher incidences of rare renal neoplasms in dosed groups are not considered to be statistically different from the control group.

2. The association is not consistent, since four out of five mouse studies did not find reproduce similar renal neoplasms at comparable doses.

3. The association is not specific, since females of this pivotal study, which have been exposed to higher levels of glyphosate did not develop renal neoplasms. Also, there were no renal findings in the LD group, whereas the control group had two.

[PAGE \ MERGEFORMAT]

Comment [wh14]: I THINK YOU SHOULD KEEP IN THE SENTENCE BELOW THAT JOHN DELETED

Comment [wh15]: I AM SUGGESTING ADDING THIS WORDING TO MORE GENERICALLY ADDRESS SOME OF JOHN'S COMMENTS THAT THE TOXICOLOGISTS AREN'T GETTING THE HILL CRITERIA

Comment [JA16]: Specificity is not considered a viable Hill principle. Consider smoking that is not specific at all – heart disease, lung cancer, (protective Parkinson's disease), oral cancer, etc.

Comment [wh17]: JOHN IS WRONG – IT IS THERE – IT IS THE 3 ONE MENTIONED BY HILL

Comment [JA18]: Strong in Hill's article refers to the size of the difference between exposure groups, not presence/absence of statistical significance

Comment [JA19]: This is not what Hill meant by specificity. He meant that the exposure only caused 1 disease. Also, specificity has been refuted as a helpful criterion – witness that smoking causes many types of cancers and other diseases

To me, this might be a matter of inconsistency unless males are particularly susceptible.

Figure 17: Continued

As the Johnson trial got closer, McClellan proposed that instead of retracting the papers, they could simply correct the conflict-of-interest section at the end of the papers and state again that the authors must declare any potential conflicts of interest. He argued that these papers might be a sensitive part of an ongoing litigation process. The back-and-forth arguments between Whalley and McClellan continued, but in the end, McClellan won, and there were no retractions. Whalley did notify the sixteen authors of the glyphosate paper of the decision to update the declarations of interest and to include what was called an "expression of concern." "We have not received an adequate explanation as to why the necessary level of transparency was not met on first submission," that statement concluded. "We thank those who brought this matter to our attention."[14]

A month later, Whalley left his post at Taylor & Francis.

By the time I arrived in San Francisco, the jury had learned about all of this: the animal studies, the toxicology data, the alleged ghostwriting issue, and how the EPA and EFSA had interpreted the evidence. They had learned how Monsanto arranged the media campaign after the IARC report became public, and how Monsanto was perfectly willing to downplay evidence that contradicted their position.

That was the context in which I was to take the stand. It was now my turn to help the jury better understand how, in my opinion, the use of Roundup had caused Lee Johnson's lymphoma. David Dickens would be guiding me through my testimony. David, Brent Wisner, and I spent several hours in the War Room going over Lee Johnson's medical records and the possible questions that Monsanto might ask me.

Brent has a boyish face but a serious demeanor. In his bio, he describes himself as being "driven by a deep-rooted passion for using the law to help those who have been marginalized and hurt by large, and sometimes malicious, corporations," and declares that he has "dedicated his practice to vindicating his client's rights and holding wrongdoers accountable."[15] And Brent was superbly prepared: he knew every study, every author, every detail, without looking at a notepad. He conveyed an impression of total confidence, as though there was no way he and his colleagues would lose this case. While some might say this level of confidence borders on arrogance, I personally welcomed it, for

I felt we needed that kind of attitude in order to win this case. I got the impression that Wisner had no reservations about pushing the jury—and pushing back against Monsanto.

The tall and trim Dickens struck me as the complete opposite of Wisner, as he was more reserved. However, as I watched the two interact, I sensed that their very different styles meant that they would make a great team in court.

When I asked Brent for his advice on my appearance the following day, he told me, "Chadi, the key to winning this trial is to be yourself. If you are, I know you'll connect with the jury. And don't forget, science is on our side." Echoing what Mike Miller had told me some months earlier, he said, "Remember that you know the science better than they do, and you know how to treat patients better than they do. Hell," he added, "George Lombardi might be 'Litigator of the Year,' but he's never treated someone with non-Hodgkin lymphoma."

After lunch at a local deli—where I avoided the french fries I craved and ordered a salad instead—Dickens suggested that we do a dry run, mimicking what would happen in the courtroom the following morning and going over his plans for my testimony. The team reminded me again that Judge Bolanos, who was presiding over this case, had denied many of the motions they'd made. For example, she had not allowed the team to bring up the state of California's Proposition 65 warning, which tells California residents that they are using a substance that "has been identified as a probable carcinogen."[16] Glyphosate had been added to the proposition list on July 7, 2017 (after Lee Johnson's diagnosis). This was another puzzling thing to me, as a legal outsider: how could we not tell the jury that the state in which this trial was taking place and where the plaintiff lived had already acknowledged the potential carcinogenicity of glyphosate but allow Monsanto to bring up the fact that Germany did not list glyphosate as a carcinogen? Germany was relevant, while California was not? I'm sure there were sound legal reasons for the judge to make those rulings, but to a layperson like me, there didn't seem to be much logic in this. But there was little use in complaining about it. We had to play the cards that we were dealt.

As usual, I could not sleep the evening before my appearance. The pressure was on again: I did not want to fail Lee Johnson, or the other patients, or for that matter the legal team that I had come to know and respect. I was fully aware that I was just one of a half dozen or so expert witnesses. But I also knew that one bad witness could derail an entire case, and I was committed to doing my best.

My experience with Griffis at the Daubert Hearing was fresh in my mind, and I spent much of the night thinking about how Monsanto would likely cross-examine me and what their strategy would be. I was certain a personal attack was heading my way and that once again my MBA degree and I would be dragged through the legal mud. But I had a few ideas on how to counter that.

"Tomorrow," I said to Monsanto in my mind, "you're going to get a little payback."

9

My Trial Testimony: *Johnson v. Monsanto*

"It's your job to just tell the facts," I said to myself in the mirror as I knotted my silver and black tie. "You know the science better than they do," I continued, as if I was giving a lecture to a class of one. "You know how to explain it. You've done this with patients and families many times before."

The face in the mirror didn't look entirely convinced. Doubts crept in again.

"I hope I don't screw up. What if my testimony derails the whole trial?" I worried. I very much wanted to do everything I could to help Lee and other patients. My concerns were always for patients and their suffering.

I knew the outcome of the trial was not only up to me. There were many more expert witnesses testifying on Lee's behalf, most of them smarter and much more accomplished than me. Still, I felt the weight of that responsibility resting on my shoulders.

I stared at my reflection in the mirror. "No, you're not going to blow this. You've prepared for it. You know your stuff. Like Mike and Brent said, you're the expert, not them. You're ready."

Pep talk over, I headed downstairs. The doorman of the Westin waved down a taxi, and I hopped into the back for the short ride from Union Square to the War Room on Van Ness.

To my surprise, I was the first to arrive. The attorneys trickled in one after another, and with them were two new faces. One, a tall man, was introduced to me as Robert Kennedy Jr., the environmental lawyer and author, son of the former presidential candidate, and JFK's nephew. That was a surprise. You don't expect to run into a Kennedy in an office in San Francisco—or most other places, for that matter. He was not involved in the court proceedings himself but is an environmental lawyer and author. At the time, Mr. Kennedy, who is graced with the distinctive family looks, was already outspoken on his anti-vax views. Views that, as a physician, I don't share. But this was 2018, two years before COVID-19, and while I was not quite sure of his affiliation with the case, or who invited him to stop by the War Room, he apparently wasn't there that morning for any other reason than to show his support for Lee Johnson and his advocates. I was certain that, as an environmental lawyer, he had little sympathy for Monsanto and their refusal to acknowledge the possible links between glyphosate and cancer.

To be honest, though, while it was certainly exciting to meet a Kennedy, I was more impressed with meeting his wife, actress Cheryl Hines, a star on one of my favorite American television shows, *Curb Your Enthusiasm*. I would have loved to talk with her about what I think is a terrifically funny show and to have told her that the show's predecessor, *Seinfeld* (co-created by *Curb Your Enthusiasm* star Larry David), had helped me acculturate to America. But of course, today there was no time for chitchat. I did, however, take a photo of the two of us, as otherwise no one would have believed I'd met her.

The courtroom was walking distance from the War Room, but if I hadn't been following the rest of the legal team, I would have had no idea where to go. There seemed to be a plethora of courthouses in San Francisco and the Bay Area. It made me think of a wryly cynical line about Los Angeles's courthouses in Michael Connelly's novel *The Lincoln Lawyer* (later made into a movie): "When the criminals get caught, they get funneled into a justice system that has more than forty courthouses spread across the county like Burger Kings ready to serve them—as in serve them up on a plate." San Francisco seemed to have an equally bewildering array of court "franchises," and that's not even taking into account those in the adjoining counties of the Bay Area. One website lists twenty-one county courthouses in Marin, Alameda, and San Francisco counties, and that's not counting superior or federal courthouses. To add to the confusion, some of these buildings have multiple names; the one where Lee Johnson's case was being heard, at 400 McAllister Street, across the street from San Francisco's historic City Hall, is known as both the San Francisco Civic Center Courthouse and the Superior Court of California–County of San Francisco. "Stone fortresses" is how Connelly described the courthouses of Los Angeles. The Civic Center Courthouse certainly fit that bill: it was a V-shaped granite building with three-story-tall windows.

Once we'd passed through security at the main entrance, it really started to sink in for me. A little over two years before, in the spring of 2016, I'd agreed to have a phone conversation with a couple of lawyers. Now it was July 20, 2018, and I was a participant in a case that had attracted worldwide notice (heck, even celebrities were here!) and in which, at least for one day, I would be the center of attention.

These proceedings would settle the question of whether Monsanto had been at fault in Lee Johnson's case. By that point, of course, I personally believed that Monsanto *was* at fault. I'd seen documented proof of the lengths to which this corporation had gone to fight what I thought was a sound conclusion that their product had caused harm to some people, including Lee Johnson.

Up until the moment I walked into the courtroom I had been fully expecting to get a call at some point from Tim or Mike or someone else saying, "Guess what? Monsanto settled. They agreed to put a warning label on Roundup." To me (and to many others) that would have been a perfectly sensible thing to do. Like the tobacco companies did with cigarettes, like alcoholic beverage manufacturers do with their products, Monsanto could have simply slapped a warning label on the container and continued selling Roundup to your average homeowner. And perhaps they could have agreed to recommend tighter safety protocols to protect workers like Lee. Case closed.

But that relatively painless solution never happened. And the fact that it didn't was even more puzzling when I learned that Monsanto was . . . well, no longer Monsanto. The chemical giant I had come to know so well had been swallowed by another, even larger entity.

When I first got involved in this litigation, I heard rumors that Bayer, the German pharmaceutical giant best known in the United States for its eponymous aspirin brand, was interested in buying Monsanto.

In September 2016, the two companies signed a binding merger agreement, which created a legal framework for the acquisition of Monsanto at a cash price of $128 per share. Bayer CEO Werner Baumann had promised that the deal would deliver "substantial value to shareholders, our customers, employees and society at large."[1] At the time, Bayer had a reported market value of about $90 billion; for Monsanto, that figure was $42 billion. As the *Wall Street Journal* wrote later, Bayer's bid for Monsanto was "designed to turn the inventor of aspirin into the world's biggest crop-science business."[2] It will undoubtedly, in my opinion, end up as a case study for future MBA students on how *not* to orchestrate a corporate merger.

At first things went well, and the acquisition made it past regulatory reviews in both the United States and Germany. And then cracks began to appear in what had looked like a slam-dunk of a deal: the first lawsuits over Roundup. But even those didn't derail the acquisition, which closed in early June 2018. Afterward, in a statement printed in trade publications from *Successful Farmer* to the *Saudi Gazette*, Baumann promised a better world through the chemistry of this merger:

> *Today is a great day: for our customers—farmers around the world whom we will be able to help secure and improve their harvests even better; for our shareholders, because this transaction has the potential to create significant value; and for consumers and broader society, because we will be even better placed to help the world's farmers grow more healthy and affordable food in a sustainable manner.*[3]

Those rosy views would soon fade faster than morning dew on the weeds Roundup was so good at eradicating. In March 2019, less than a year after the merger, *Barron's* would opine, "It's looking as if Bayer's acquisition of Monsanto is a candidate for the pantheon of truly terrible mergers-and-acquisitions deals. . . . Like many disasters, this one has an air of cursed inevitability."[4] And that was hardly a minority opinion. In August 2019, the *Wall Street Journal* would write, "Bayer's $63 billion dollar gambit ranks as one of the worst corporate deals in recent memory—and is threatening the 156-year-old company's future."[5]

As the number of lawsuits increased, shareholders and major investors started wondering about the due diligence that Bayer had supposedly conducted ahead of the acquisition. I was curious about that as well. Why would Bayer move to pay billions of dollars to acquire a company that was plagued with lawsuits? According to a Bloomberg Law analysis of the deal, Baumann and his legal team "had been aware of controversy around glyphosate but banked on a lack of scientific evidence establishing Roundup's risks."[6]

I often wondered what Bayer executives back in Germany must have been saying behind closed doors as the trials progressed. Suddenly it was their name that was being dragged through the mud. Suddenly it was Bayer that was being linked to products causing cancer. If nothing else, it was a public relations disaster. And it wasn't lost on the shareholders: in the April 2019 annual shareholder meeting, usually a rubber-stamp occasion, senior management lost a vote of confidence, as 55 percent of investors expressed their mistrust in the CEO—something that Bloomberg called a "bombshell."[7] Reportedly, Baumann survived the purge with the backing of Bayer's chairman, Werner Wenning. But clearly the message from Bayer's shareholders and the investment community was: "What the hell were you thinking?"

Years after the initial announcement, Bayer agreed to strengthen external oversight of its due diligence in dealmaking. And Bayer hired an independent lawyer to review the legal advice it had commissioned before the Monsanto acquisition. To no one's surprise, this "voluntary" special audit came back with favorable results—that is, Bayer had conducted the proper due diligence prior to the deal and there was nothing suspicious to be found.

Really?

It's unlikely that we'll ever know the exact details of what Bayer knew and didn't know before it acquired Monsanto, but there will always be the speculation that had they done their homework, they would have not pursued a deal that now seems destined to go down as one of the worst in modern corporate history. I always wondered whether Bayer would ever admit to themselves that they made a mistake. I hate to be cynical, but I suspect the answer to that question is "Of course not."

Then again, my job wasn't to second-guess high-level corporate strategy. Nor was it up to me to assess blame. That would be the responsibility of the jurors in the Monsanto trials.

In the San Francisco courtroom that day in July 2018, I studied the Johnson trial jurors as they took their seats. These seven men and five women had been selected after careful questioning by attorneys for both sides. I'm sure that some of them would have identified as Democrats, others as Republicans. Could the former be more sympathetic to the plaintiff, while the latter might lean toward the defense—toward Monsanto? I hoped the jurors wouldn't make a decision in Johnson's case solely because of their political leanings; at least from my vantage point that day, these people seemed earnest and intent on understanding the often complex and emotionally charged issues in this case the best they could, and with an open mind.

My role in this had been made clear to me: as an expert witness, I needed to make these good citizens see our side of the argument, to see the science without getting them bogged down in too many scientific details. This jury had already heard the testimony of several other expert witnesses retained by the plaintiff attorneys. Almost unavoidably, some of that testimony had been technical, deep, and complicated. I had to remind myself that I was here, in part, to bridge the gap between the experts and an average person. Like many physicians, I must play a similar role when interpreting medical information to our patients and their families. When I was about to explain the results of a test or a treatment protocol, I always needed to ask myself: "Could somebody be confused by what I'm going to tell them? Could it be interpreted another way?" Same thing here. Clarity and simplicity are key. At this point, it's about using the evidence and the data to win over the jury and have them see it your way—which, in this case, I was convinced was the right way.

There were about fifty people dispersed on both sides of the courtroom. I suspected that there were journalists, bloggers, and maybe a couple of law students among those in attendance. I wondered whether the seating represented their loyalties: were those who supported Lee on one side, while those backing Monsanto sat on the other?

I have given many scientific presentations over the years and have lectured in front of large and small audiences. For me, as for most experienced speakers, there is always an initial sense of anticipation, even fear, when I step up to the lectern. But once I start talking, my heart rate slows down and my posture relaxes. I was hoping that the same thing would happen here.

"You may call your next witness," Judge Bolanos said to the plaintiff counsel.

David Dickens stood up. "I now call Dr. Chadi Nabhan to the stand."

I felt fifty pairs of eyes on me as I stood and walked to the witness stand. I glanced over at Judge Suzanne Ramos Bolanos. She wore dark-framed glasses, and as I took my seat, she was studying papers in front of her. She spoke softly, and I did not see her make eye contact with the jury very often.

A California native, Judge Bolanos had gotten her law degree from Yale and had an impressive record that included considerable experience in social justice organizations, such as the Mexican-American Legal Defense Fund, where she'd begun her career.

After I lifted my right hand and swore accuracy and truth in my testimony, David started the direct examination by asking me a few personal questions. He had advised me the night before on the importance of the jury getting to know me as a person, not just as an expert physician, and so that's how we started.

David Dickens: *Can you please introduce yourself to the jury and tell them a little about yourself?*

Chadi Nabhan: *Sure. My name is Chadi Nabhan. I am a hematologist and medical oncologist. I have been so for the past twenty years. I live in Chicago in the northern suburbs. I have twin boys turning eleven, going on twenty-five. So, I am sure you can understand the challenges.*

I heard a few chuckles from the jury.

DD: *It's probably good to get at least a little break from them; is that fair?*
CN: *Yes.*

Under David's guidance, I briefly described how I'd come to the United States from Syria after graduating from medical school.

DD: *I want to run through your educational background. You received your medical degree in 1991, is that correct?*
CN: *Yes, in 1991. And this was followed by two years of basic research at Mass General Hospital and Harvard Medical School in Boston. That's why I'm a Patriots fan. [Pause] Now I've lost everybody in the courtroom.*

Another laugh from the jury box. I was starting to relax a little; I wasn't trying to be entertaining, but simply being myself. My sometimes-dark humor is well known to my friends and colleagues, most of whom think I am not funny; I wanted the jury to feel like they knew me and could connect with who I am.

I told them about how I'd worked for a year after my residency as a primary care doctor in an underserved suburb on the South Side of Chicago. I then started my fellowship training in hematology and medical oncology at Northwestern University's Robert H. Lurie Comprehensive Cancer Center. It was there that I started becoming more interested in blood cancers, such as leukemias and lymphomas, rather than other types of cancers. I worked diligently in the lab run by the cancer center's director and began studying lymphoma. I was doing research, seeing outpatients and inpatients, and publishing scientific articles. In short, I was gradually becoming an expert in this form of cancer.

This led to Dickens asking me about my research. We reviewed for the jury several papers I had written on lymphoma. While I understood the need to establish my bona fides to the audience, this felt a little like bragging, and I'm not someone who would normally brag about my experience or my research accomplishments. I never felt there was anything to brag about, to start with. My guiding principle has always been that I should always do more, achieve more, prove more. This feeling intensified as I got older. It's as if I'm in a race, but I don't know whom I'm racing against and don't know where the finish line is.

But the attorneys had told me the previous day, "This is not the time to be modest or humble." They wanted the jury to understand my qualifications and to establish a trust factor so that they would see my opinions and recommendations as valid and credible.

So, gently prodded by David, I shared with the jury how I'd joined a large hospital system, Advocate Health Care. I became chief of oncology and director of the cancer institute at one of their flagship hospitals, Lutheran General, in Park Ridge, a suburb of Chicago. I then joined the University of Chicago as the medical director of the Clinical Cancer Center and cancer clinics. I described to them my administrative, clinical, and re-search responsibilities. Dickens made sure that the jury knew that I am licensed in five states, including California. He asked me if I was allowed to practice medicine in California if I wanted to, and I answered affirmatively. Like a proud father, Dickens added to the jury that I am board certified in internal medicine, hematology, and medical oncology. "Board certified in *three* disciplines of medicine," he said to the jury. "Did I get that right, doctor?"

I nodded uncomfortably.

Yet while I wanted to move on from the topic of my credentials, it was important for him to ask me about my employment at the time. The jury had to understand not only my experience but how what I do now compared with what I did earlier in my career—particularly because this was surely going to be seized upon and twisted out of shape by the Monsanto side. We continued to follow my career path: after leaving the University of Chicago in the summer of 2016, I joined Cardinal Health as a vice president and chief medical officer of their Specialty Solutions division, providing medical oversight to six business units including regulatory sciences, real-world evidence, research and publications, and relations between oncology drug manufacturers and practices that were using Cardinal Health as a distributor. I chose this route after receiving my MBA in health-care management and having gained more in-depth knowledge about healthcare finance and economics, as I felt that there was more that I could do to help patients with cancer in my new role. I hoped that this resonated with the jury. But to make sure the jury didn't think of me as an out-of-touch corporate suit and to reinforce my bona fides as a researcher, Dickens reminded them that I had authored more than three hundred peer-reviewed manuscripts, abstracts, and book chapters, and that I was continuing to publish and do research.

David's direct examination of me was the first part of what's called voir dire, the process by which someone is qualified as an expert witness. At this point in the proceedings, the Monsanto attorney would have been allowed to object to my testifying as an expert. Lombardi, the Monsanto counsel, did not, somewhat to my surprise; I wondered if this was a strategic move, or maybe it was a sign of confidence on his part that, regardless of my qualifications and expertise, he would win the showdown.

Dickens then started asking me about Lee Johnson's medical condition as well as his prognosis and the treatments he had received. I answered his questions by first explaining to the jury in broad terms what cancer is: essentially an uncontrollable growth of cells that results from an imbalance between cell growth and cell death. This imbalance can occur for a variety of reasons, including exposure to something in the environment. But cancer is not one single disease. Treatment and prognosis often depend on the organ the cancerous cells originated in. It was also critical for me to highlight to the jury the complexity of non-Hodgkin lymphoma classifications and how these have changed over the years. I pointed out that the latest iteration of these classifications, published in 2016, identified more than sixty subtypes of non-Hodgkin lymphoma.[8] I felt the jurors needed to realize that the epidemiological research couldn't possibly look at every single one of these variations of the disease, so although of course we were guided by the findings from these studies, we had to keep in mind that each form of lymphoma had its own characteristics. This discussion also allowed me to clarify why older epidemiological studies were unable to explore the relationship between every subclassification of lymphoma and Roundup: the numbers of each lymphoma subtype in these studies were generally too small to make a statistically significant conclusion about every single subtype.

Dickens then asked me to perform a "differential diagnosis." In legal terms, this is a process by which the expert reviews all potential causative and risk factors of a patient's condition, then goes through each one by one until, through process of elimination, we're left with the likely culprit. This is rather different from how we physicians define "differential diagnosis," which is considering various possible diagnoses that might have similar symptoms and then reaching the proper diagnosis. A better term for what I was asked to do on the stand might be "differential etiology," since the diagnosis is already known and what I was trying to do was identify a possible cause.

I asked the judge if I could step out of the witness box, and she nodded. I walked to a whiteboard that was set up facing the jury. With a dry-erase marker, I started writing down risk factors for non-Hodgkin lymphoma. As I scribbled, I apologized for my often-illegible writing: "My handwriting is very poor. You know us doctors." I noticed some smiles on the faces intently watching me.

At the top of the board, I wrote "Non-Hodgkin Lymphoma: Known Causes." Under that I wrote every potential risk factor that I thought was possibly implicated in non-Hodgkin lymphoma:

- Age
- Sex
- Race
- Family history [of hematologic malignancies]
- Viral and bacterial associations
- Use of drugs that suppress the immune system
- Autoimmune disease
- Pesticides/occupational exposures

Just to be inclusive, I added tobacco and alcohol to the list, as well as sun exposure. As I'd said earlier in one of my depositions, to my knowledge there is no conclusive and convincing data that any of those last three factors cause lymphoma, but I knew that the defense would bring up these factors as possible confounders, so addressing them head-on was a good idea.

I then went down the list of risk factors one by one and explained how each could be eliminated.

"Mr. Johnson is younger than the median age of patients who are generally diagnosed with this disease, so we can strike that," I said, striking through the word "age." I tried to make eye contact with each and every member of the jury. One young woman sitting in the front row was taking copious notes and nodding, which made me think I was gaining momentum. The others also seemed intent and focused on what I was saying.

I went through a few more. "Sure, lymphoma is more common in men than women," I explained. "But this does not mean that being male in and of itself causes lymphoma." I drew another line through the word "sex."

I then explained that certain cancers can be more prevalent in certain races or ethnicities. In fact, most types of lymphomas are more common among whites, and Lee Johnson was African American. But in my opinion, race is not a causative factor in lymphoma. I looked the jury in the eyes as I struck out the word "race."

By the time I was done ruling out possible causes, what remained on the whiteboard was the word "pesticides," under which I'd written "glyphosate." I hoped that would leave an impression with the jury.

We addressed the chronology of Lee's case. It was critical to explain to the jury that Lee had been exposed heavily to Ranger Pro for two years before he was diagnosed with T-cell lymphoma. This led Dickens and me to discuss the latency period of the disease. It was an opportunity for me to clarify to the jury the concept of latency, and to show that latency isn't a fixed length of time but depends on the severity and chronicity of exposure to a toxic compound. As I've explained previously, we can't say for certain that in order to develop cancer from exposure to some substance, a person has to have been exposed for more than a year, or twenty years, or whatever number you want to choose. Lymphomas can develop in their own time, at their own pace. And this is where clinical judgment comes in. This is where you need an experienced oncologist to look at the patient, to look at all the evidence, and—drawing upon what we know the literature tells us, but also being cognizant of what it doesn't tell us—make a judgment about the likely cause.

When the judge announced a thirty-minute break, I noticed several of the jurors looking intently at me and at the whiteboard. They seemed to be fully engaged in what I was saying. A good sign, I hoped.

David certainly felt so. "It's going according to plan," he said to me.

A few minutes into the break, I heard a murmur go through the audience and turned my head to see Lee Johnson himself entering the courtroom. He looked better physically than I had expected, but he showed little emotion. I'm sure the whole setting was as unnerving to him as it was to me. He took a seat beside Brent, David, and the other lawyers. I went over to him and, recalling his sensitivity to shaking hands, just smiled. "Good to see you," I said. He smiled back tersely and nodded. I wondered if he'd had to be coaxed to show up; I was sure it had to do with how he was feeling that day. When you're going through the kind of treatment Lee was, you take things one day at a time.

Regardless, seeing Lee motivated me. It really is all about the patients. They are the ones who are enduring the illnesses. They're the ones whose lives are on the line. And, in cases like Lee's, they're the ones who were never warned that a chemical they were working with could have this sort of impact on them.

Remembering that what I and the other experts were doing was about Lee added to my determination. I felt that so far, I had acquitted myself pretty well as David guided me through the key points of my testimony—but soon the questioning would become much less friendly. Having Lee in the courtroom focused me and strengthened my resolve to do my part to make sure justice would be served.

When we resumed, Dickens finished his direct examination, with me offering that in my opinion, Lee would have been unlikely to have developed this lymphoma if it wasn't for Ranger Pro. He then announced to the judge, "Your honor, I pass on the witness." This was the signal that it was now time for the other side to get their shot at the witness.

Lombardi rose. In his arms were two large white folders full of papers and deposition transcripts. He carefully laid them in front of me on the bench. These were the papers he'd be referring to during his questioning.

I must admit I was somewhat intimidated by the sheer size of these folders. I was also intimidated by Lombardi. Even before he said a word, I could tell he had total command of the courtroom.

The Litigator of the Year smiled at the jury, politely thanked me, and then immediately went on the attack. He questioned the fact that I was not currently seeing patients in clinical practice. He then asked about where I worked and the revenue of the company that employs me, as if this had anything to do with the case. David's objections to the relevancy of this line of questioning were denied.

Lombardi's style was interesting. I felt that his tone toward me was both accusatory and condescending—and yet at the end of each question he looked over at the jurors with a gentle, reassuring smile, as if to say, "I'm really just asking what needs to be asked here. I'm not a bad guy."

After reminding the jury that I worked at Cardinal Health Specialty Solutions as the chief medical officer, he proceeded:

George Lombardi: *Cardinal Health is a huge company; right?*
Chadi Nabhan: *It is.*
GL: *And your division is a huge division, isn't it?*
CN: *Yes.*
GL: *And is it, what . . . an $11 billion division; is that right?*
CN: *You're talking gross revenue?*
GL: *Yeah.*
CN: *Yes. That's actually significantly less than Monsanto.*
GL: *Thank you for that, Doctor. I hadn't actually asked you that. But you don't dispute that Cardinal Health is a Fortune 15 company?*
CN: *Yes; it's number fourteen, according to their last list.*

The tone and entire exchange were familiar. As Kirby Griffis had done earlier, the Monsanto attorneys were trying to make the jury think that I could not be trusted because I had joined a successful healthcare company with good revenue. I had hoped the jury saw through this approach.

GL: *You haven't seen patients since you started at Cardinal Health?*
CN: *That is correct. And it's actually by design—*
GL: *Thank you, sir. You answered my question. I just asked you whether you see patients anymore.*
CN: *Is it okay if you don't interrupt me, please?*
GL: *Sir, you need to answer my questions, not go beyond my questions.*
CN: *But I prefer not to be interrupted.*

Lombardi then went on to discuss the epidemiological studies that I had reviewed. He focused on the fact that these studies failed to adjust for exposure to other pesticides. Interestingly, when we discussed the paper from 2003 where the authors did indeed adjust for over forty other pesticides, Lombardi downplayed the type of mathematical model that the study's authors used, logistic regression, and argued that another type of mathematical modeling, hierarchical regression, was what counted.[9] Rather interesting, given that just a few minutes earlier he had been arguing that the other studies we had been discussing should have used logistic regression. I explained to Lombardi that clinicians commonly use logistic regression models, not hierarchical regression models.

"Have you polled them, all clinicians?" he challenged.

"I can guarantee it to you," I shot back.

When he introduced the AHS paper, which is the study that Monsanto relied on the most, I saw an opportunity to point out the inconsistency of Monsanto's argument and their selective use of studies:

GL: *I'm not going to go into detail here, Doctor, but you know the results of this [AHS] study, right?*

CN: *Yes, although the details are important to interpret. And you went through the details of every other study. It's only fair to go through the details of this one.*

Out of the corner of my eye, I thought I saw David Dickens smile and nod. Point for our side.

We went on to discuss how the AHS had been updated in 2018 from the initial report published in 2005. He seemed to want to breeze right past it, but I felt compelled to ask him a question.

CN: *So, are we not going to discuss the limitations of the 2018 study, as we did with the earlier one?*

GL: *If we have time, we may, Doctor. I have limited time.*

CN: *For the jury, you just discussed the limitations of other studies, and this study has many limitations, I think it's only fair to discuss that.*

At that point, I'd been on the stand for over two hours and my throat was getting dry. I asked for water, and Lombardi—no doubt looking to remind the jurors what a nice guy he was—reached over to his table and handed me a sealed bottle of water.

I took it and made a face. "Any glyphosate in this?"

The entire courtroom broke into laughter.

If the jury laughs, I had been told, it means they like you. Maybe that had gotten under Lombardi's skin. I hoped it had.

He then asked me why I hadn't added the term "idiopathic" to the "differential diagnosis" I wrote on the whiteboard. I countered that I had indeed told the jury that most non-Hodgkin lymphomas are idiopathic by nature, meaning they are of unknown cause; I reminded him that what I had written on the whiteboard were the *known* causes. I thought it was implied that the term "idiopathic" is applied only when we fail to identify any known etiologic cause. Lombardi insisted that I should have kept idiopathy in the differential.

I turned my head to the jury. "Just to be clear, I was discussing the known causes." I saw a few heads nod in sympathetic agreement.

I don't think that Lombardi's focus on the use of the term "idiopathic"—essentially, the "we don't know what causes non-Hodgkin lymphoma" argument—won him points in the minds of the jury. The logic of his argument, in my view, was strange, and yet Monsanto would continue to use it in future depositions and trials. Essentially, they argued that the patient could have developed the idiopathic type of lymphoma and that there was no way to tell with 100 percent certainty whether that particular person's disease had a known cause or an unknown cause.

I did my best to explain that we call a disease idiopathic *only* when our efforts to identify a cause fail. If we were to believe Monsanto's proposition, then we would have to assume that a heart attack in a heavy smoker had nothing to do with smoking, or that a diagnosis of type 2 diabetes in an obese person is never related to their obesity.

Meanwhile, I was cognizant that while we were throwing this medical term around, I wasn't sure it had been clearly defined to the jury. I turned to them and said, "By the way, you might be wondering about this word 'idiopathic.' It's a term physicians use to sound smart when they don't have an answer. It basically means 'we don't know.'"

Lombardi continued his line of attack, questioning my reliance on the IARC report and challenging the epidemiological studies that had shown association and causation between glyphosate and non-Hodgkin lymphoma. He insisted that there were many confounding factors that could cloud the interpretation of these studies, but I argued that this is the nature of epidemiology. There are no perfect studies; every study can be critiqued and challenged. Still, we need to consider how these studies affect the way we treat individual patients.

Lombardi tried to ridicule the IARC as an organization by suggesting that it had previously proposed that coffee causes cancer. I smiled when I heard this, because I knew about this research. I explained to him and to the jury that very hot drinks have been associated with esophageal and stomach cancers, especially in Eastern cultures, where people tend to consume more of these very hot drinks. I explained that it's not the coffee or tea per se but rather the high temperature of these drinks that could be problematic. "So don't worry about Starbucks and Dunkin' Donuts," I said.

Lombardi then shifted gears and tried to challenge the chronology of events. A point of contention in the trial was how long it took Lee to develop lymphoma after his exposure to glyphosate. The shorter the time, the better for their argument, as a shorter period would raise the likelihood that it was caused by something else. But in truth, it took about two years of him spraying five days a week before he started seeing symptoms. That's plenty of time for causation, in my opinion.

Lombardi then took another tack, arguing that none of Lee's treating physicians had suggested that glyphosate was the cause of Lee's lymphoma. I explained that unless they had done research on the topic, they wouldn't know about the link. "I wasn't aware of these associations until I was asked to get involved in this case," I admitted. I turned in my chair to face the jury box, which was on my right. "I'd just like to point out that the American Cancer Society has listed pesticides as a major risk factor for developing lymphoma, and Lee was exposed to only one pesticide. How can that be ignored?"

Lombardi kept on with this line of questioning, pointing out that one of the physicians who'd treated Lee was at Stanford, and that physician had never suggested that glyphosate had anything to do with Lee's lymphoma. "Stanford is one of the finest universities in the world," I acknowledged, mindful that the university's Palo Alto campus was only about a thirty-five-minute drive away from the courtroom. "But last I checked, that doesn't mean that every physician affiliated with Stanford is right every time. No physician is."

As Lombardi was approaching the end of his cross-examination, he went for the kill: he tried to impeach me by suggesting that I had lied under oath. I don't think he succeeded, but I will concede he scored a hit at the very end. It was over the question of whether Lee Johnson could have developed his particular form of lymphoma, mycosis fungoides, without exposure to glyphosate.

GL: So, Mr. Johnson could well be someone who would have developed mycosis fungoides when he did, whether he was exposed to glyphosate or not?

CN: I don't believe so. I do not believe so.

GL: Okay. Let's go to your deposition on January 30, 2018.

My heart sank. Why was he asking me to look back at my prior deposition? And why was he grinning like the cat that ate the canary? My concern heightened when Dickens objected and asked the judge for a sidebar conversation. But he was overruled, again.

Lombardi resumed his questioning. With a smirk, he asked me to look at my January 2018 deposition.

GL: And this is your testimony under oath at your deposition. You're familiar with that process, obviously?

CN: Yes.

GL: Same oath that you took before you testified today; right?

CN: Can't play crystal ball with patients developing cancer or not, true.

GL: And let's look at slide 7.

Lombardi showed on the screen a similar question I had been asked back in January 2018: "Mr. Johnson could well be someone who could have developed mycosis fungoides when he did, whether he was exposed to glyphosate or not for all you know, correct?" Then he proclaimed: "Your answer, under oath, was 'He could have,' isn't that correct?"

"We can't really tell with one hundred percent certainty if someone is going to develop cancer today or tomorrow or never," I said.

"Thank you, Doctor," he said dismissively. We both knew he'd won that round.

With that, my time on the stand ended. I was exhausted. Despite that last-minute score by Lombardi, I thought I'd stood up well overall.

* * *

I had booked a flight back to Chicago that same evening, and Tim Litzenburg drove me to the airport after I was done.

"You did well today, Chadi," he said.

"Eh," I replied. "I don't know."

"We never know until the verdict is read. But you went toe to toe against one of the best litigators in the country and held your own." Seeing that I was still brooding over the final exchange with Lombardi, he added, "We all want to see justice done for Lee. But this isn't like a boxing match that ends in a knockout. It's going to come down to who can hang in there and win on points. You got us some points today."

At the airport, I threw my entire two hundred pounds onto the first chair I saw in the United Club lounge and tried to relax and unwind. But my mind kept replaying every exchange that Lombardi and I had had. I knew that I had done what I could, and that the verdict usually reflects the sum of what the jury believes from all witnesses, not just one.

As I stared at the darkening skies outside, I couldn't help thinking that if Lee lost his case, it would be my fault.

It was a long flight home.

10

The Johnson Verdict

After I returned to Chicago, I continued to follow the trial proceedings from afar, via social media, newspapers, press accounts, and of course communicating with the legal team.

I was particularly interested in the testimony of Dr. Timothy Kuzel, the Monsanto expert witness who was to oppose my views from a medical oncology standpoint. I have known Tim for a long time: he was a faculty member at Northwestern when I was a fellow in training there. I also had seen him at various conferences and meetings. At the time of his testimony, he was the chief of hematology and oncology at Rush University Medical Center, a prominent academic institution in downtown Chicago. Kuzel has a specific interest in the form of lymphoma that Lee suffered from. He also has interests in immunotherapy.

I have always respected Kuzel's opinions, although we clearly were on different sides in this litigation battle. It was rather interesting, I thought, that the expert oncologists for the plaintiff and defense were from Chicago and knew each other.

On direct examination, Kirby Griffis highlighted Kuzel's expertise in lymphoma and in mycosis fungoides. I would never question Kuzel's expertise in this specialty; I learned much from him and others when I was at Northwestern. But he admitted himself that he was not an expert in epidemiology—a critical gap in terms of his expertise in this case, as the data linking glyphosate to non-Hodgkin lymphoma involves being able to interpret the mountains of epidemiological literature that I had forged through during my research.

Interestingly, Kuzel also acknowledged to the jury that the slides being shown on his behalf in the courtroom had been created by Monsanto. He was quick to add that he agreed with their content, however.

While I had stated that most non-Hodgkin lymphoma cases are idiopathic, Kuzel, during his testimony, asserted that *all* cases of mycosis fungoides are idiopathic and refused to acknowledge that there could be any possible cause. Kuzel also contradicted me and suggested that Johnson's initial rash and symptoms started in the fall of 2013, shortening the latency period, which is a point that Monsanto had wanted to impress upon the jury. If the latency was short, then Monsanto could make an argument that glyphosate did not have enough time to cause the lymphoma.

Kirby Griffis: *Now, Dr. Nabhan testified, sir, that the majority of non-Hodgkin's lymphoma is idiopathic, meaning of an unknown cause. Do you agree with that?*

Timothy Kuzel: *Yeah. I think that's true.*

KG: *And what about mycosis fungoides?*

TK: *That's true.*

KG: *And would you use a stronger word than the "majority" for mycosis fungoides?*

TK: *I would say every case of mycosis fungoides is of unknown etiology.*

KG: So, if you have something that is majority idiopathic or 100 percent idiopathic, is there any way to rule out idiopathic when you're evaluating cause, in your opinion?

TK: You can't rule out idiopathic unless you can, with absolute certainty, pin things down. I don't tell every lung cancer patient that I encounter, even if they smoked, that cigarette smoking is the cause of their lung cancer, because there are lung cancers which arise in nonsmokers. There's always the possibility that it was something else. Unless there's clear, scientific facts, there's no way to determine what caused any patient's cancer.

"Whoa," I thought when I heard that. Certainly, some lung cancers occur in nonsmokers, but to tell a smoker that their lung cancer was not related to smoking is a totally different story. While some might call it semantics, words matter in the courtroom: I knew the jury would pick up on it. And of course, so did David Dickens, who capitalized on this later during his cross-examination.

Dickens was masterly in how he conducted his questioning, making sure that the jury knew something else about Kuzel that wouldn't stand in his favor there: the fact that he admitted to having not reviewed all the body of research—"the literature," as it's called in scientific and academic publishing—on Roundup and non-Hodgkin lymphoma.

David Dickens: Now, Doctor, you agree mycosis fungoides is non-Hodgkin lymphoma; we can agree on that?

Timothy Kuzel: Absolutely.

DD: And once again, your opinion in this case is specific to the question of whether or not glyphosate can cause mycosis fungoides; correct?

TK: Yes.

DD: You didn't look at anything with respect to non-Hodgkin lymphoma?

TK: That's correct.

DD: You didn't look at epidemiology of non-Hodgkin lymphoma generally?

TK: Only in the setting of some of the recent epidemiologic work that I think we brought up earlier in the agricultural worker survey which was more focused on that.

DD: That's the Agricultural Health Study you're referring to; correct?

TK: Yes.

DD: And you're aware that it found a quadrupling of the risk of T-cell lymphoma?

TK: It didn't.

DD: You say that because it's not statistically significant; is that the reasoning?

TK: Yes. There was a wide range of possible impacts on the diagnosis.

DD: So, we'll get to that later, but that's the study you reviewed; correct?

TK: Yes, regarding more global non-Hodgkin lymphoma.

DD: And you didn't do a literature search on your own in this case?

TK: No.

I was surprised to hear that Monsanto's expert witness admitted that he had not reviewed all the epidemiology literature on Roundup and non-Hodgkin lymphoma. He was basing his judgments on a single study, the AHS.

"That's a win for us," I thought.

Dickens jumped on this:

DD: *So literally, the only case you reviewed with respect to Roundup glyphosate and non-Hodgkin lymphoma—the only case you reviewed was the Agricultural Health Study?*

TK: *Yes.*

DD: *The case that Monsanto claims is the biggest and the best?*

TK: *I think regardless of who claims is biggest or best, I reviewed the study.*

DD: *And that was, once again, provided to you by Monsanto?*

TK: *Yes.*

DD: *Did they provide you any other epidemiological studies on the question of whether or not glyphosate or Roundup can cause non-Hodgkin lymphoma?*

TK: *No.*

DD: *And you didn't go out and do your own literature search to find additional studies; did you?*

TK: *I was not looking for causes of non-Hodgkin lymphoma.*

Dickens decided now was the time to remind the jury about what Kuzel had said about smoking during the questioning by Kirby Griffis.

DD: *You said that even if you were treating a patient with lung cancer who smoked, you wouldn't tell them that smoking was the cause? That was your testimony; correct?*

TK: *That's correct.*

DD: *So, is it fair to say that before you would give any positive causation opinions, you'd have to be 100 percent sure?*

TK: *No. I will often tell a patient I believe it was cigarette smoking in the case of lung cancer. But if they ask me if I am absolutely certain, I tell them no, because there are other causes of lung cancer.*

DD: *Okay. So, you tell them, "It's not certain, but it could have substantially contributed to your lung cancer."*

TK: *In the case of cigarette smoking, yes.*

Most of us can agree that lung cancer in a smoker is attributable to smoking. Certainly, I would imagine the members of the jury did—and I bet they were beginning to feel as if they were having smoke blown in their faces by Monsanto's arguments.

Dickens won points on that, I'm sure, but I think the knockout blow came when he asked Kuzel if he had reviewed a PET scan Lee Johnson had done in June 2018, the month before the trial started. That scan showed without a doubt that there had been significant progression in Johnson's disease. Kuzel had stated earlier on direct examination that Johnson was in remission, citing a PET scan done in March 2018.

DD: You're aware that Mr. Johnson had a recent June 2018 scan demonstrating a progressive disease of cutaneous lymphoma scattered throughout his body?

TK: Well, I'm aware that he had a PET scan that showed some uptake in a number of areas.

DD: Okay. And what was your reading of that PET scan?

TK: It was suspicious for recurrence.

DD: Okay. And so, did you reach an opinion whether or not it was recurrence?

TK: No, because I didn't have any physician exam information to correlate those findings with.

DD: So, even though you reviewed that scan, you still sat here and testified to the jury that he was in complete remission?

TK: No, I think he may be relapsing.

Dickens then referred to data on Lee from March 2018 that Kuzel said he had read, and which was included on a timeline he had created, tracking the course of Lee's disease, that had been presented to the jury earlier.

DD: *All right. Well, your timeline—*
TK [interrupting]: *Ended March 2018.*
DD: *Okay. And so you didn't mention the new PET scan, did you?*
TK: *Not on the timeline, no.*
DD: *Why didn't you put it on there?*
TK: *Because I'm not sure what it means.*
DD: *You don't believe he has progressive disease?*
TK: *I am suspicious that he has progressive cancer.*

There was nothing more that Dickens needed to say after that, and nothing more Kuzel could say. Even Monsanto did not ask their expert any further questions. I speculated that they were glad his testimony was over.

Closing arguments started on August 7, 2018. Brent Wisner made it a point to reiterate that the principal issue of the trial was not cancer, or the influence of big corporations, but choice. In a concise and simple manner, Wisner summarized animal studies, information about the ways that glyphosate could cause cellular damage, and epidemiological studies. He then explained the importance of choice:

> *You see, Monsanto made a choice when it got Dr. Parry's report in 2000 to not share it with the EPA or anyone. They made a choice to not do the studies that Dr. Parry recommended, because they knew what it would show. They made a choice to not study the formulated product in long-term carcinogenicity studies. And the best explanation I could hear so far was that's because it would have killed all the rats and mice. They made a choice when they didn't call Mr. Johnson back, when he called the company desperate for answers, asking for some information about the stuff that he was spraying at the time. And when Dr. Goldstein didn't call him back, that continued to deprive him of that choice. Monsanto made a choice to engage in ghostwriting. You've seen document after*

document where Monsanto's response to scientific concerns, legitimate
scientific concerns, is to make up science.

In a brilliant move, he then extended the theme to build a bridge to the jury by linking the concept of choice to the individual decisions members of the jury made during the trial itself:

I know you guys didn't actually have a choice to be on this jury, so it's kind of a weird thing to thank you for your service, but . . . you've asked incredibly good questions. Some of them we were able to answer. Some of them we were not. But the questions told us exactly how closely you were tracking this case. Some of you have five notebooks of notes. That's unbelievable. The level with which you've paid attention to this case, thank you, and I really mean that.

He reminded the jury that Monsanto had not brought any of their employees to court to look the jury in the eye and defend the company. "They sent surrogates," he said, with contempt in his voice.

He then went on to hammer home that Timothy Kuzel, the expert oncologist retained by Monsanto, had based his testimony on one piece of evidence: the AHS, a study that was provided to him by Monsanto. Wisner contrasted that with my review and how I conducted a comprehensive analysis of all epidemiological literature that was available. Based on that evaluation, he said, I had concluded that Roundup caused Lee Johnson's lymphoma.

Wisner reminded the jury that at one point, when Johnson's treatment choices were being discussed, Kuzel had suggested that Lee might benefit from bone marrow transplantation. And then Wisner hit back hard: "That Monsanto would call someone up here and speculate about bone marrow transplants that no one has ever offered to him, when his most recent scan showed [that his disease was progressing], is outrageous. It is disgusting. It is reprehensible. That man has no dignity. I'm thankful I wasn't here for that. I was writing a brief in the back room for most of it. When I was reading the transcripts, I turned red."

I felt bad for Kuzel, who does have dignity and is not reprehensible; he is a good physician and a scientist. Wisner was a bit harsh with these comments, I thought. Still, it was a big error on the defense's part, and because of what was at stake here, Wisner had to capitalize on it. It's what lawyers do.

Wisner told the jury that they needed to send a message to Monsanto so that the company could learn a lesson in humility. Near the end of his closing statement, Wisner gestured toward one of the attorneys sitting with the Monsanto team.

Brent Wisner: *On her cell phone is a speed-dial to a conference room in St. Louis, Missouri. And in that conference room, in that board room, there's a bunch of executives waiting for the phone to ring. Behind them is a bunch of champagne on ice.*

George Lombardi: *Your Honor, I object. This is supposed to be about the evidence. This is complete fantasy.*

Judge Suzanne Ramos Bolanos: *Sustained.*

BW: *The number that you have to come out with is the number that tells those people, "We have to change what we're doing." Because if the number comes out and it's not significant enough, champagne corks will pop. "Attaboys" are everywhere.*

GL: *Your Honor, it's the same objection.*

SRB: *Sustained. Mr. Wisner, please do not engage in speculation. You may continue.*

Brent acceded to the judge's decision and went on with his closing argument. But he had managed to create a vivid image in the minds of the jury: an image of arrogant corporate fat cats sitting around their executive suite, chuckling at the inevitability of them again getting what they wanted, confident that they could buy, bluster, or intimidate their way out of this problem, just as they had done with the science around glyphosate many times before.

But they were not quite confident enough to actually show up in the courtroom and look Lee Johnson or the jurors in the eye. You think Wisner was going to let that little detail pass?

"Where are the Monsanto employees?" Brent thundered. "I mean, think about this for a second. They're being sued. I think we have a valid case here. And they didn't bring a single human being from the company to look into your eyes and say why they did what they did. They didn't bring a single live human being here to talk to you, talk straight, and say, 'Here, this is what we thought. This is why we did it.' Why? Why didn't they bring anybody?"

There were, Brent continued, only two possibilities. "One, they couldn't find somebody who could do that and not commit perjury. Or, two, they were afraid of all the documents . . . showing that there's no conceivable way they didn't know the risk."

With that, he urged the jury not to give Monsanto a reason to celebrate, and finished his performance in bravura fashion:

Today, in this room, tomorrow in deliberation, and when you return a verdict, we're going to make it right. And your verdict will be heard around the world. And Monsanto will have to finally do something—conduct those studies they never conducted and warn those people they never warned. Thank you for your time.

"Wow," I thought. "This guy was born to litigate."

Brent asked the jury for over $370 million in punitive damages. He justified the amount by adding up the interest Monsanto would have collected since 2012 (the year Johnson started spraying) on their yearly revenue.

Lombardi then delivered the defense's closing argument. I'm certain that his courteous manners and gentle smile appealed to the jury. His arguments? Maybe not quite as much.

One of Monsanto's main contentions was, essentially, that because the EPA said Roundup was not carcinogenic, it could not possibly be carcinogenic. Monsanto wanted everyone to ignore all the other available data and assume that the EPA was infallible and impervious to potential political pressure. Lombardi noted that Roundup had been on the market for over forty years and claimed that there were "countless" studies that had never shown any potential carcinogenicity; he conveniently forgot the many studies we had highlighted, not to mention the false IBT registration data and dubious ghostwritten research, both of which had been highlighted during the trial.

"Forty years of this product on the market," he said confidently. "Forty years of this product being regulated. Forty years of scientific studies ranging from human to animal to cell. The evidence is clear. The message from that evidence is clear, and it's that this cancer was not caused by Ranger Pro."

It was interesting how Monsanto argued that animal studies should be dismissed because the doses used in animals were not similar to the doses humans were typically exposed to. Monsanto's argument centered on the idea that glyphosate had adverse effects in animals because the doses studied in animals were higher than what humans might get exposed to. This argument, as plaintiff experts pointed out, was not valid, as animal studies by design use higher doses so that experiments can provide results rapidly.

Predictably, Lombardi pointed again and again to the AHS, the one study that Monsanto loved. He went on to praise Dr. Kuzel, attributing his lack of knowledge about the progression of Lee Johnson's disease to a simple misunderstanding. And he repeated that Johnson's treating physicians had never stated in the medical records that Roundup caused Johnson's lymphoma.

He then moved on to discredit the expert witnesses who had testified for the plaintiff, including me. He described me as a "retired practicing doctor," and throughout his closing argument repeatedly referred to me as a "former doctor." He told the jury that I was an executive who works for a Fortune 500 healthcare company—as if that was inappropriate. A particularly strange argument, I thought, considering that he was representing what was now one of the largest healthcare and agribusiness corporations in the world.

He also highlighted that I was the only physician in the courtroom to have testified that glyphosate caused Johnson's lymphoma. "If it were that easy to figure out the cause of mycosis fungoides, why didn't we figure it out a long time ago? If it were easy, if Dr. Nabhan is actually the guy, this would be a huge medical accomplishment, discovering the cause of mycosis fungoides, the first person in the world to do that."

Lombardi even defended the fact that Dr. Goldstein, the Monsanto medical director, who had never called Lee Johnson back when his symptoms were just appearing and he needed help, saying that it had no impact on the medical case. "How about common courtesy or compassion?" I thought angrily when I heard that. "That should have been sufficient reason to call him back." And I didn't think that would go over well with the jury, either.

All that said, I think Lombardi lived up to his superstar reputation and was able to plant doubts in the jury's mind.

In his rebuttal, however, Brent came back with both barrels blazing. As noted earlier, Lombardi had brought up the point again that Lee's treating physicians never mentioned glyphosate as a possible cause of his lymphoma. Brent reminded the jury why that was easily explained, referencing my testimony that before 2016 I had been unaware of the relationship between Roundup and non-Hodgkin lymphoma, and had come to the conclusion that they were linked only after I had spent many months researching the topic. Similarly, there was no reason that Johnson's doctors would have reached that conclusion either unless they too had carved out hours from their busy schedules to take deep, time-consuming dives into the literature.

And speaking of those physicians, in another deft twist Wisner drew the jury's attention to the fact that Monsanto failed to have any of Johnson's doctors appear in this courthouse to testify—and these physicians were nearby, at Stanford and University of California, San Francisco. Unlike many experts who had traveled from all over the country to be present, they hadn't made the thirty-minute drive from Palo Alto or the twenty-minute drive from the UCSF campus. Why not? Why weren't they testifying? Again, I thought, the Monsanto strategy of keeping key people out of the courtroom seemed to be backfiring on them.

Brent also hit back at Kuzel's testimony, reiterating that Monsanto's oncology expert relied only on one epidemiological study. But Kuzel's testimony was damaged even more by having said that Johnson was in remission, despite the evidence to the contrary that had been presented during the trial.

I was grateful that Brent went on to refute Lombardi's arguments against me and the other expert witnesses. He also told the jury that Monsanto hid behind the EPA because they had a cozy relationship with the agency.

Once the closing arguments were done, Judge Bolanos provided her instructions to the jury, advising that they could ask questions via notes they could pass to the judge at any time during deliberations and could review transcripts of any testimony from any expert who had appeared in court, as well as the depositions by other witnesses who did not testify in person. The

Figure 18: Judge Bolanos reading the Johnson verdict on August 10, 2018.
Source: Reuters/Alamy Stock Photo

jury also was advised they could have physical access to any
document, photo, or item entered into evidence. Finally, the
judge provided instructions on potential awards should they find
Monsanto at fault.

The jury deliberated for three days. On August 10, 2018, Judge
Bolanos announced that a verdict had been reached. (See figure 18.)
Back in Illinois, I was driving back home from work when I learned
that a verdict had been reached; I pulled over to a side street and
waited to hear news on the result by clicking on social media outlets and YouTube channels.

The courtroom was packed with journalists and others who
fought for seats, eager to witness the historic verdict. You could
imagine the media frenzy, with everyone scrambling to hear
the outcome of the first Roundup trial. Lee Johnson was present, of course, though his wife was not; she had to work in order to feed the family that Lee, because of his illness, could no

longer support—a point that had been duly made by Brent and David. And later I found out that, interestingly, a few Monsanto employees had somehow managed to slip into the back of the courtroom.

At 2:15 p.m. local time, Judge Bolanos entered the courtroom. It was silent; everyone waited to hear what she would say. A verdict had been reached, she announced. And then, rather than revealing it right away, she said that it would be read in a larger courtroom upstairs, to accommodate the crowd and media. During the few minutes' delay, my knuckles were turning white as I grasped the steering wheel of my car and waited for the reading of the verdict to resume.

Finally, everything was ready, and the twelve jurors entered, with the others in the room rising out of respect. I was certain that everyone was trying to guess the verdict by reading the jurors' facial expressions.

"Has the jury reached a verdict?" the judge asked. They acknowledged that they had. A folder that contained what the entire world was waiting to know was handed to the bailiff and then to the Judge.

Judge Bolanos opened the envelope and read the verdict silently. Then she read each of the questions and answers aloud.

1. *Are the Roundup Pro or Ranger Pro products ones about which an ordinary consumer can form reasonable minimum safety expectations?* **Answer: Yes.**

2. *Did Roundup Pro or Ranger Pro fail to perform as safely as an ordinary consumer would have expected when used or misused in an intended or reasonably foreseeable way?* **Answer: Yes.**

3. *Was the Roundup Pro or Ranger Pro design a substantial factor in harming Mr. Johnson?* **Answer: Yes.**

4. *Did Roundup Pro or Ranger Pro have potential risks that were known or knowable in light of the scientific knowledge that was generally accepted in the scientific community at the time of their manufacture, distribution, or sale?*   **Answer: Yes.**

5. *Did the potential risks of Roundup Pro or Ranger Pro present a substantial danger to persons using or misusing Roundup Pro or Ranger Pro in an intended or reasonably foreseeable way?*

   Answer: Yes.

6. *Would ordinary consumers have recognized the potential risks?*

   Answer: No.

7. *Did Monsanto fail to adequately warn of the potential risks?*

   Answer: Yes.

8. *Was the lack of sufficient warnings a substantial factor in causing harm to Mr. Johnson?*   **Answer: Yes.**

9. *Did Monsanto know or should it reasonably have known that Roundup Pro or Ranger Pro were dangerous or were likely to be dangerous when used or misused in a reasonably foreseeable manner?*   **Answer: Yes.**

10. *Did Monsanto know or should it reasonably have known that users would not realize the danger?*   **Answer: Yes.**

11. *Did Monsanto fail to adequately warn of the danger or instruct on the safe use of Roundup Pro or Ranger Pro?*   **Answer: Yes.**

12. *Would a reasonable manufacturer, distributor, or seller under the same or similar circumstances have warned of the danger or instructed on the safe use of Roundup Pro or Ranger Pro?*

    Answer: Yes.

13. *Was Monsanto's failure to warn a substantial factor in causing harm to Mr. Johnson?*   **Answer: Yes.**

As I heard each of those yeses, the smile on my face got wider. Despite the sometimes stilted wording of the questions, the jury's verdict was clear: Unanimously, they had found that Lee's cancer was related to his exposure to glyphosate. Unanimously, they had found Monsanto guilty of malice—a legal term that means the intention, without justification or excuse, to commit an act that is unlawful.

Then came the moment when the judge announced the damages that the jury had awarded—exactly what Monsanto would have to pay Lee Johnson.

14. *What are Mr. Johnson's damages?*   **Answer:**

Past economic loss:	Future economic loss:
$819,882.32	$1,433,327.00
Past noneconomic loss:	Future noneconomic loss:
$4,000,000	$33,000,000

15. *Did you find by clear and convincing evidence that Monsanto acted with malice or oppression in the conduct upon which you base your finding of liability in favor of Mr. Johnson?*

Answer: Yes.

16. *Was the conduct constituting malice or oppression committed, ratified, or authorized by one or more officers, directors, or managing agents of Monsanto acting on behalf of Monsanto?*

Answer: Yes.

17. *What amount of punitive damages, if any, do you award to Mr. Johnson?*   **Answer: $250,000,000.**

This was a legal slam dunk! An emotional Lee hugged his lawyers (see figure 19). And two thousand miles away, sitting in my car, I said aloud to the jury, "Thank you," even though they couldn't hear me.

As thrilled as I was, part of me was also a bit surprised. I knew we had a good case; I knew we were on the right side. But all too often, being right, being innocent, does not guarantee what happens in a courtroom or a tribunal. There are many places in this world where the verdict for a trial like this would have been a foregone conclusion, because the powerful always win over the powerless. But here, in this courtroom, in this country, in this particular case, the so-called little guy toppled the giant. David beat Goliath. There was justice at the end.

Oh, of course Monsanto would continue to deny any wrongdoing, and would appeal the decision, but that did not matter. The guilty verdict had been announced to the world.

Later that day, Lee Johnson and his legal team held a press conference (see figure 20), and I was able to watch some of it. Johnson was graceful and thanked Tim Litzenburg for having been by his side throughout the proceedings. He also thanked Mike Miller for his support.

Figure 19: Lee Johnson hugging David Dickens after the verdict was announced on August 10, 2018.

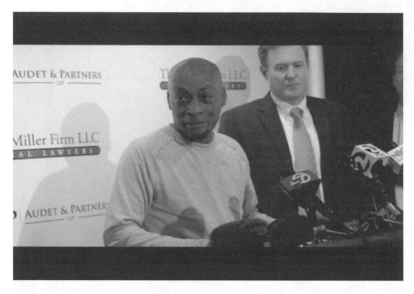

Figure 20: Lee Johnson at his press conference after the verdict, with David Dickens in the background.

"This case has been very hard," Lee said to the sea of note-pads and microphones in front of him. "It's been fought long. It took a lot of time to get here, and these guys did a lot of work, and I am very proud of each of you."

Amen to that, I thought. The legal team deserved so much credit for assembling all the needed evidence, for developing an effective litigation strategy, and for supporting Lee throughout this ordeal.

Then it was Wisner's turn to speak to the press. "Every major known carcinogen had a moment like this," he proclaimed, sounding as if he was gearing up for another closing statement. "A moment when the science finally caught up. When they could no longer bury it. Where people had to actually look at it and say, 'We have a problem.' And this case is that moment. Because right now a unanimous jury here in San Francisco has told Monsanto, 'Enough. You did something wrong and now

Figure 21: Chadi Nabhan with the lawyers after his testimony on July 22, 2018. (*Left to right in the back row*) Jeff Travers, Mark Burton, David Dickens, Richard Mayer; (*front row*) Pedram Esfandiary, Robert F. Kennedy, Jr., Brent Wisner, Timothy Litzenburg, Dewayne "Lee" Johnson, Chadi Nabhan, Kathryn Forgie, Michael L. Baum, Kevin Baum.

you have to pay.' . . . We now have a way forward. I want to say thank you. This was a massive team effort." Indeed, that team eventually won the "Trial Team of the Year 2019" award from the National Trial Lawyers Top 100 outlet (see figure 21).

That evening, I had a very good night's sleep.

You would think that this would have been the end of the Lee Johnson case. But it wasn't, and I must admit that I find this particular aspect of the American legal system rather puzzling. Follow my logic, and perhaps you may feel the same.

A trial is held that follows all the proper procedures.

A verdict is decided upon by the jury.

A judge pronounces the verdict. Unless there's going to be an appeal (meaning, in essence, a new trial), that should be the end of the story—or at least that chapter, right?

Apparently not.

A month after the historic verdict, in September 2018, Monsanto filed not only a motion for a new trial, but also a motion for something called a JNOV. I had to look that up; it is a legal acronym for "judgment notwithstanding the verdict." Essentially, Monsanto was requesting that the judge nullify the verdict rendered by the jury should the judge find that "there was no factual basis for the verdict, or it was contrary to law."[1] This made no sense to me; wasn't trial by jury the whole idea here?

In their request for this "do-over," Monsanto's attorneys rehashed the same arguments they'd been making all along: the EPA, the supposed lack of causative scientific evidence, the AHS. On top of that, they added a new and rather creative one. I call it "the champagne defense." In addition to claiming that the basic facts of our argument—that exposure to glyphosate caused Lee's illness—were wrong, they criticized Brent Wisner's passionate plea during closing arguments, and in particular his memorable line about how Monsanto's executives were back in St. Louis, gathered in a conference room, with the champagne corks ready to pop if the verdict went their way. The defense team argued that it might have swayed the jury against Monsanto.

To me, this whole thing sounded like they wanted to override the jury verdict because they'd been outlawyered. This now seemed less an argument over substance and more like injured pride—or maybe just desperation. "Do something!" might have been the panicked call they got from St. Louis after the jury verdict left the champagne bottles securely corked, and this was the best the defense attorneys could come up with. In fairness, the plaintiff side might have done the same had they lost.

While I know that filing a motion for a JNOV was perfectly legal, the whole idea of it stuck in my craw. Running to the judge after the verdict sounded like bad sportsmanship. It was as if at the end of a football game, when the teams are heading back to the locker room, some guy from the losing team instead decides to pick up the ball and run down the empty field, expecting the unopposed touchdown to count. Of course, this is how I saw it as a citizen and a legal outsider.

My incredulity at this tactic only grew when I learned that *I* was part of the JNOV motion. Monsanto specifically asked the judge to entirely ignore my opinion. Why? Because, Monsanto claimed, I had completely overlooked the possibility that Lee Johnson might have had an idiopathic lymphoma. I had not even brought up such a possibility, they argued. This was an interesting and creative twist on my testimony, since I had clearly explained to the jury, while in front of the whiteboard, that doctors use the term "idiopathic" only when they fail to find a cause for an illness.

It stands to reason that if a risk factor is identified and is a known causative risk factor for the disease in question, then the term "idiopathic"—meaning "we don't know"—can no longer be valid. But in Lee Johnson's case, we can be reasonably sure we *do* know. Take a different medical example. Could we tell a heavy smoker that their lung cancer is idiopathic? If someone who has uncontrolled high blood pressure and is obese suffers a heart attack, could we tell them and their family that we have not the foggiest notion of what might have caused it? It seems clear to me that we *cannot* say that; we cannot merely shrug and say, "we don't know." While we don't have ironclad proof that years of smoking was the reason for that individual's lung cancer, we have sufficient evidence to suggest direct causation. While we don't have reams of data allowing us to say with 100 percent certainty that the patient's heart attack was caused by their high blood pressure and obesity, we certainly have a basis for saying those are likely to have been substantial contributors. That was absolutely the case here, I thought. Monsanto seemed to be grasping at legal straws with this almost semantic argument.

Nonetheless, the motion was made, and a response was required. On October 2, 2018, Lee Johnson's legal team filed two briefs in opposition, one arguing against a new trial and the other against the JNOV motion. They rebutted every argument Monsanto made. "The trial court ensured that Monsanto received a fair trial from a fair and impartial jury," read the brief, which was written collectively by the team. The jury was "excellent," paid careful attention, sifted through the evidence carefully, and came to an appropriate verdict. Monsanto's motion for a new trial should be denied. And they concluded their opposition to the JNOV motion with what I thought was a good, hardhitting summary:

> *Monsanto fails to meet its heavy burden of overturning a unanimous jury verdict. . . . Attacks on the admissibility of plaintiff's experts cannot challenge the jury's finding of the sufficiency of the evidence based on the competent testimony presented by plaintiff's experts. Moreover, Monsanto's rehashed arguments that regulatory approvals of Roundup shield it from liability are still meritless. . . . It is time for this company to compensate Mr. Johnson for the tragic harms incurred by him. For all of the foregoing reasons, Monsanto's motion should be denied.*

A little over a week later, on October 10, the two teams appeared in front of Judge Bolanos for oral arguments so that she could decide on next steps. Mike Miller was there, having recovered from his boating accident, and was leading the arguments in front of Judge Bolanos against Monsanto's motions.

One of the questions, as I've mentioned, was whether my testimony should be excluded. I must say that I was insulted; why was I part of this process? The answer, as Mike explained to me later, was that I was the "case-specific" expert who had said that Lee Johnson's cancer was caused by Monsanto's product.

Back in Chicago, my ears should have been burning that day, as Lombardi actually suggested in front of Judge Bolanos that my testimony was "outrageous" and "invalid" because I had never explicitly stated that Johnson's disease could have been of unknown cause. "Outrageous"? "Invalid"? I got angrier and angrier when I heard what Monsanto's lawyers had said.

I was grateful to learn that Mike Miller tenaciously defended my testimony to Judge Bolanos. But perhaps even more decisive in what happened next was due to the presence that day in the courtroom of several jurors from the Johnson trial. I imagine that as they sat there listening to Monsanto's counsel essentially arguing that their verdict should be overturned, they probably felt a little like I did—insulted that their efforts, and by implication their intelligence, were being questioned by Monsanto. They were also very concerned about what they'd heard during the debate over the JNOV. They recognized that Lombardi, in particular, had not painted an accurate picture of what was said during the trial.

The following day, one of them—Juror #4, Robert Howard—sent Judge Bolanos a letter. Two other members of the jury would write similar and equally powerful letters. But Howard's was the first and is particularly eloquent, as you can see in this slightly abridged version:

October 11, 2018
Dear Judge Bolanos,

It is with all due respect that I have to say that what transpired at [the JNOV hearing] was astonishing for several reasons. I feel it is my civic duty to address these reasons with Her Honor.

First, on the differential analysis of Dr. Nabhan. The fact of the idiopathy of NHL was made abundantly clear by more than one witness.

Secondly, all parties agreed that the epidemiological leg of the tripod of causation was weaker than the other two, but a tripod with one weak leg stands nonetheless. Again, common sense.

Third, the whole discussion of non-economic damages was an embarrassment to the humanity of anybody who was there, except perhaps Monsanto. To alter this award on a technical issue would be a travesty. Common sense and decency.

Fourth, your instructions to the jury were crystal clear. We had them in writing and referred to them often. On the matter of Mr. Wisner's opening and closing, they were colorful, but I had already disregarded what he said by the time I got to the jury deliberation room, as I had Mr. Lombardi's opening and closing. Simply following your instructions.

Finally. . . . You instructed that we could assess the credibility of witnesses and discount their testimony in whole or in part. . . . To say that Monsanto employees, and at least one expert witness, were clearly uncomfortable would, in my opinion, be a vast understatement.

All parties agree that we were an exceptional jury. We were, in fact, praised by Monsanto, and you, Your Honor, for our attention, intelligent questions, etc. Yet our integrity, intelligence, and common sense has been cleverly and openly attacked by inference. The idiopathy issue, the science, the non-economic damages issue, what is and what is not evidence, the higher bar for punitive damages, credibility of witnesses: I got it. I believe my fellow jurors got it. With all due respect, Your Honor, I don't see how this can go both ways. Monsanto can't ask for a jury, state that we intelligently and with diligence considered only the testimony and evidence, and methodically weighed that evidence— and then turn around and infer that we must have ignored your instructions and did not comprehend the evidence! It just doesn't add up.

The possibility that, after our studious attention to the presentation of evidence, our adherence to your instructions, and several days of careful deliberations, our unanimous verdict could be summarily overturned demeans our system of justice and shakes my confidence in that system.

I urge you to reconsider your tentative ruling and to not completely overturn the punitive damages and I also urge you to leave the liability intact.

Respectfully,
Robert Howard, Juror #4

Hearing about these letters almost provoked in me the same tearful reaction I'd had when I heard the verdict. In particular, the words of Robert Howard—a man I never personally met, but will always be grateful to—were moving and reaffirming.

Six days after Mr. Howard wrote that letter, on October 22, Judge Bolanos issued an order denying Monsanto the JNOV. But she also lowered the total award to Johnson to $78 million. I knew that Miller and his team would reluctantly agree to the lower award, as they were shifting their focus to future trials. To fight it would have meant a retrial. Nobody wanted that.

And so, with a rap of Judge Bolanos's gavel, this legal encore was complete. The way I look at it, the group that had insisted on coming back out and playing one more song that no one wanted to hear had been booed off the stage.

There would still be a formal appeal, which would also be denied, but for all practical purposes, the historic Lee Johnson trial was done.

11

Hardeman

On February 4, 2019, I strode up the walkway to the Phillip Burton Federal Building in San Francisco, where I had endured the Daubert hearing the year before—and where I'd been adroitly skewered by Kirby Griffis and left Judge Chhabria unimpressed. I was back to help prove again that Roundup was linked to another specific case of non-Hodgkin lymphoma, this one suffered by a man named Edwin Hardeman.

I had been introduced by Tim Litzenburg to Kathryn Forgie, a partner, at the time, in the law firm Andrus Wagstaff, one of the four major legal firms in the litigations against Monsanto and the one handling Hardeman's lawsuit. Kathryn had reached out to Tim because she'd been in the courtroom during my testimony in Lee Johnson's case and thought I could be helpful in this one.

A sharp attorney, Kathryn has laser-focused eyes and a deep reservoir of compassion. We would become good friends over the course of this trial. Earlier in her legal career, Kathryn had defended pharmaceutical companies, but she now argues from the opposite point of view, and I think it's because she has such a humane streak in her—she came to feel the way I did about the battle we were fighting and decided that if she was going to get involved, she should do so on the side of the underdogs.

On our first call, she addressed me as "Dr. Nabhan."

"Please call me Chadi," I responded. "No need for the 'Doctor' stuff. Otherwise, I need to call you 'Counsel' all the time."

She laughed. "Deal!"

Her first request was for me to give her a primer on non-Hodgkin lymphoma. As she put it, "Whatever you would tell a first-year medical student—or a first grader—will be fine." She had a sense of humor.

We also had a common interest—but divided loyalties—in another area: soccer. Kathryn is an avid supporter of the English team Chelsea, while yours truly has been a fan of Manchester United since 1981, well before I moved to the United States and developed a passion for American football (I am still proud of being that rare fan who loves *both* forms of football).

After reviewing his medical records before agreeing to serve as an expert witness in the Hardeman case, I dived back into the literature. I read the latest glyphosate research; there had been a new batch of studies, as interest was growing in the scientific community, sparked in part by the publicity the lawsuits were getting. I studied further Hardeman's medical records and familiarized myself with his condition and the treatments he'd undergone.

Soon after I started my research, I got a package in the mail from Kathryn. More documents? No—she had sent me a Chelsea soccer jersey for good luck. I found that very funny, as there is no love lost between Manchester United fans and Chelsea fans. But that is how Kathryn operates, and she made it way more fun to be part of the team despite the seriousness of the matter. And there would be no shortage of serious challenges ahead of me.

The Hardeman case was part of the MDL—the multidistrict litigation—that represented all consolidated federal cases that were to be tried in front of Judge Chhabria. While Johnson's trial had been held in state court, Hardeman's was going to be held under federal jurisdiction, and his was the first of these federal trials to go to litigation in front of a jury. What happened here could influence the thousands of other Roundup cases that were filed or expected to be filed in the coming months.

You may be wondering, like I did, why some cases were in federal court while others were in state courts. It's complex, and lawyers often argue over which venue a trial should be held in. When it comes to Roundup-related litigation, plaintiffs can sue Monsanto either in state court in Missouri (because Monsanto's headquarters are there) or in a court in the state where they were injured. If a plaintiff elects to sue Monsanto in Missouri, the case must stay in state court, and Monsanto cannot successfully argue that the case should instead be tried in a federal court in the district that encompasses Missouri. But if a person injured in, say, California sues only Monsanto, and no other defendant, in California state court, Monsanto has the option, under what's known as the "removal statute," of arguing that the case should be moved to federal court.[1] In the Johnson trial, a pesticide distributor located in California was initially also named as a defendant, and so the case had to stay in state court. Ultimately, the case against the distributor was dropped.

Chhabria also decided that the Hardeman trial would be bifurcated, or divided, into two parts. The first phase would focus on whether glyphosate can cause lymphoma in general and in Mr. Hardeman's case in particular. If the first phase of the trial found that it could, then the second phase would focus on damages and liabilities, including punitive ones. But when I initially heard that the Hardeman case would involve relitigating the issue of whether glyphosate can cause lymphoma, I thought, "Again? Haven't we been down this road before?" I envisioned another trial going back and forth about the IARC and the AHS, the mouse studies, the Parry study, blah blah blah.

I suspected Monsanto was happy about this, and it turned out they were—the plaintiff attorneys had argued against bifurcation unsuccessfully.

Edwin Hardeman had had extensive exposure to Roundup. As he and his wife, Mary, would tell the *Guardian* later, he had regularly sprayed the herbicide on his properties for nearly three decades—first at his home in the California coastal town of Gualala, and later on at his fifty-six-acre property in Santa Rosa, where he used it to control poison oak. He said it had never occurred to him that he was putting his health at risk: "It's a product that is so widely used. It's in all the home building stores. There's no warning label."[2] Like with Lee Johnson, this seemed to me like another example of a hardworking man who had put his faith in a widely available product, but then developed lymphoma as a result.

Kathryn asked me to write a formal expert report that would be shared with Monsanto prior to my deposition in this case. The report was to summarize how I approached the case, by looking at all risk factors for non-Hodgkin lymphoma and then excluding the factors that were not applicable.

In reviewing Hardeman's records, I learned that he had a history of hepatitis C that had been diagnosed many years prior to his lymphoma. It had been treated effectively in 2005, a decade before the lymphoma was diagnosed in 2015. While hepatitis C *is* a risk factor for non-Hodgkin lymphoma, everything I had learned about that virus suggested that the risk exists for patients who have active disease, not a disease that has been treated successfully, and especially if that treatment was almost a decade prior. In fact, since November 2006 when his doctors checked, there had been no evidence of the virus in his blood, and even when he was diagnosed with the lymphoma, tests failed to reveal any evidence of liver disease or damage. I explained in my report that the lack of hepatitis C in his blood for so long prior to his diagnosis made this virus a negligible risk factor for this patient's lymphoma, and I supported my opinion by citing several papers published in the literature. I knew, however, that this point would be a contentious one with Monsanto, who seemed to believe that every single risk factor known to humankind can cause lymphoma . . . *except* Roundup.

While Hardeman was over sixty years old, I argued that age by itself does not cause lymphoma. That was important, for the majority of cancer diagnoses and deaths are in patients over sixty-five. If we were to blame age for every cancer or even every lymphoma in older patients, then we would never bother to investigate the cause of cancer in this group of people. I had a particular interest in older patients with cancer and had designed several lymphoma studies in geriatric patients, so this detail was critical to me. If we investigate causation in younger patients but dismiss every case of lymphoma or other cancer in people over sixty-five as being caused by their age, that's discrimination against older patients. Moreover, it's bad science, or maybe lazy science—and it certainly would provide a very convenient out for Monsanto.

Hardeman had presented with swollen lymph nodes in December 2014; he had these biopsied in early 2015, and a diagnosis of diffuse large B-cell lymphoma (DLBCL) was confirmed. This is another type of non-Hodgkin lymphoma, different from what Lee Johnson had. It's the most commonly diagnosed lymphoma: 30–35 percent of all diagnosed lymphomas are of the large B-cell type. It is considered an aggressive cancer, fatal if not treated properly and in a timely manner. Hardeman received six cycles of chemotherapy; thankfully, he responded very well to treatment and achieved a complete remission. That doesn't mean it was easy, a point we should never forget. Indeed, in the *Guardian* story, Mary Hardeman hinted at how hard it was for her husband and expressed bitterness at Monsanto's seeming indifference to their suffering: "They should have been with us when we were in the chemo ward."[3]

Another aspect that was important for me to address in my expert report was obesity. Mr. Hardeman's body mass index (BMI, a measure calculated on the basis of weight and height) was 32, which is considered a marker of obesity. The minute Monsanto lawyers saw that, they would be chomping at the bit, arguing that obesity was the major risk factor in this patient's lymphoma. There have been some studies that linked obesity to lymphoma and other cancers, and there have been some speculations as to how obesity might contribute to the development of malignancies, but none of these studies were conclusive or confirmatory.[4] Furthermore, we would have to take into account other issues such as someone's weight gain or loss over time, and the connection between when a person was obese and when their lymphoma developed. All of this is to say that there are many reasons why we cannot simply say, in my opinion, that obesity causes non-Hodgkin lymphoma. I have always counseled patients about weight loss, diet, exercise, and staying fit, but I have never told a patient that his or her lymphoma was caused by obesity. I explained in my report that even if there were studies that suggested a minor link, it was my belief that obesity contributed little if any to lymphoma development.

Still, on the surface, the Hardeman case appeared favorable to Monsanto. Since Hardeman was overweight, was older, and had had hepatitis C years earlier, I'm sure in the minds of the Monsanto lawyers this added up to one thing: a verdict favoring Monsanto.

After reviewing thousands of pages of his medical records, I met Edwin Hardeman in my office in Chicago on November 16, 2018, a few weeks prior to submitting my report and being deposed. I found him pleasant and energetic, and he answered all the questions I asked in the course of taking a detailed medical history, including about his hepatitis C. We then discussed his chemotherapy experience and the side effects that he'd encountered, which included peripheral neuropathy (tingling and numbness in the fingers and toes) and fatigue. Chemo can suppress white blood cell counts, which makes patients much more vulnerable to infection; we often try to counteract this effect by giving patients injections of growth factor to help keep the white blood cell count high, but Edwin also had one of the side effects of such a treatment, which is severe bone pain.

I wanted to use my time with him to better understand his exposure to Roundup. It's not unusual for patients not to remember the exact details of how, when, and where they sprayed Roundup; after all, people were told that Roundup was safe and posed no danger, so why would they recall every detail associated with using such a product? Assume I asked you how many cups of tea you had last year. You won't be able to remember the exact number, but you'd say something like, "I drank a lot of tea, maybe once a day except weekends." However, Monsanto always tried to pounce on any vagueness like that in patients' testimonies. Edwin told me that he had been exposed to Roundup since 1988 and typically had sprayed it biweekly, a couple of hours at a time, using a hand sprayer. When I asked Edwin if he'd worn protective gear while using it, he looked at me, smiled, and asked, "Why would I? I was told it's safe."

Not every case I was involved in went to trial, but I was designated as an expert in several other cases, and so I was deposed multiple times. Why some cases never make it to court while others do is beyond my understanding, but as we all know, some cases do settle out of court while others get dropped by the plaintiffs.

The week before Thanksgiving 2018, I was deposed in the case of a woman in her late thirties who was diagnosed with an aggressive case of non-Hodgkin lymphoma after two decades of spraying Roundup. Kathryn Forgie was the patient's attorney, and in the course of working with her I learned a lot. "Depositions are not memory tests," she told me one night over dinner, for example. "You can ask to see any paper and any record to refresh your recollection." Still, I continued to prepare for depositions as if I were studying for an exam. Maybe that's why I was always tense before the three major trials that I appeared in. And, indeed, sometimes a defense attorney would ask a question pertaining to a manuscript or a publication where I didn't recall all the details, and I would follow Kathryn's advice and ask for a copy of the paper so that I could review it. Oftentimes the attorney would say, "Take all the time you need to review," and I always wanted to respond to that by saying, "Well, great! Come back in about an hour!" And that was just for papers I'd already read; sometimes a lawyer would thrust a study I'd never seen before in my face and ask my opinion. At such moments I always felt that everyone in the room was watching me, even timing me, and was bothered if I took more than a minute or two, though of course it isn't realistic to review and digest technical information so quickly.

I also learned the hard way how often lawyers twist research findings out of context, in the effort to get an expert witness to say something that they could repurpose later as a sound bite in order to make the witness's statements appear inconsistent. And they'd draw on any information they could in order to find some grounds for claiming inconsistency.

A surprising and disconcerting realization was that Monsanto tried to learn as much as possible about their opposing expert witnesses, as one could never predict when such knowledge would come in handy. When I arrived for the deposition in this woman's case, the Monsanto lawyer mentioned that he was a fan of the Giants, the New York football team, and that he was ecstatic when the Patriots lost. At first that seemed like a harmless comment, an attempt to feign cordiality before he went for my jugular. But then it hit me: how did he know that I was a Patriots fan? Right there and then I realized that the Monsanto team knew about me more than I could have ever imagined. They must be following my tweets and social media posts—and who knows what else? If I posted something that might help their cause during the trial, you bet it would be used. To be fair, I'm almost certain that the plaintiff attorneys were doing the same for Monsanto's experts.

During the actual deposition, this attorney asked me about my methodology and how I had reached the conclusion that Roundup caused the woman's lymphoma. I explained that in any malignancy, including lymphoma, we would list all potential causative factors and then proceed with a process of elimination—and I reminded the defense lawyer that I had discussed the same topic during the Johnson trial, when I'd written causative factors on the whiteboard in front of the jury. He countered by asking about the "error rate" of such methodology. Even the premise of such a question is incorrect; I told the opposing counsel that this question is akin to him asking about the error rate of using a stethoscope while conducting a physical examination on a patient.

There was a lot of back-and-forth during this deposition, with the lawyers on both sides making many objections and interruptions, and sometimes engaging in heated exchanges with the opposing attorney. I learned that when lawyers keep asking the same question but in a different format, it meant that I was on the right track—they didn't like my answers, so their strategy was to keep asking and rephrasing in the hopes that I might change my answer. Little did they know that when I am right and confident of my answer, I won't change my mind or what I say, no matter how many times they rephrase the question.

A lot of what I'd learned during that November deposition came up again during my deposition for the Hardeman trial, which took place in a conference room at a Marriott near Chicago's O'Hare Airport.

Just before the deposition started, the lawyers on both teams exchanged pleasantries as if they were old friends, buddies. I wondered how they could be so friendly to each other even though they'd be at each other's throats during the proceedings; all I could figure is that it's part of the game.

After I was sworn in, a Monsanto attorney I was meeting for the first time, Brian Stekloff, started by wanting to know exactly how much money I was making for my role in this litigation. I almost laughed—how predictable! He probably spent twenty minutes on this, including asking what percentage of my total income now came from my work as an expert witness. Then he moved on to other topics, including one that I was well prepared for: Edwin Hardeman's hepatitis C. I had argued that Hardeman had indeed had hepatitis C, but it had been properly treated and, in my opinion, did not cause his lymphoma. To refute me, Monsanto cited a report by their own expert, Dr. Alexandra Levine, who was at the City of Hope National Medical Center and specializes in HIV-related cancers. Dr. Levine had opined in her report that Hardeman's hepatitis C had likely caused his lymphoma, even though it had not been active for years prior to his cancer diagnosis. She supported her opinion by citing literature that claimed hepatic cellular damage could last even if the infection was treated successfully. Stekloff threw all of Dr. Levine's findings at me. I responded with my opinion, which was that, with all due respect to Dr. Levine, I did not think hepatitis C was the culprit here. In my opinion, cellular damage to the liver from hepatitis C required the virus to be active, and since in Edwin Hardeman's case it had been treated successfully before any damage had occurred (judging by tests that assessed liver function), I did not think that hepatitis C was the issue here.[5]

While the deposition was short, and I think I acquitted myself well, this only reinforced to me that this question about hepatitis

C would be vigorously debated in Judge Chhabria's court. I welcomed that debate.

Furthermore, I suspected that Stekloff did not want to show all his cards during the deposition, which lasted barely two hours. He would certainly go on the attack in front of Judge Chhabria, so why waste time and effort here?

Another day, another flight to San Francisco. It was February 3, 2019, Super Bowl Sunday, and my beloved Patriots were playing the Rams for the championship. I got a morning flight from O'Hare, and that night, after meeting with Kathryn and the rest of the legal team, I was able to watch the fourth quarter of the game and savor the Pats' victory.

As I always did before a deposition, I worried late into the night. I still had the sense that Judge Chhabria hadn't liked me at the Daubert hearing prior to the Johnson trial. I thought I'd seen looks of impatience, frustration, even disdain on the judge's face during parts of my testimony; was that accurate, or had I been imagining it? I didn't know, and it was bothering me. And I was aware that, because from a scientific standpoint the core dispute was going to be rather simple—Monsanto would claim that Hardeman's prior exposure to hepatitis C had caused the lymphoma—the Monsanto lawyers were going to try to prevent me from presenting my opinion by attacking my methodology and then asserting that I should not be allowed to testify in the trial.

The next morning, February 4, I headed into Judge Chhabria's courtroom for another Daubert hearing focused on the Hardeman case. In a departure from the usual pattern of court proceedings, in which I would be questioned first by the plaintiff lawyers, who might offer me few softball questions, Judge Chhabria ordered that I undergo cross-examination first. That meant I'd experience tough questioning right off the bat. I was not sure why the judge wanted to do it this way, but I reminded myself of something Kathryn Forgie had said during one of our preparatory meetings: "Trials are won on cross-examination."

I was to be questioned by the attorney who had deposed me two months earlier, Brian Stekloff, a brainy Georgetown University Law School grad who had co-founded a firm in Washington, DC, and was a serious player in this kind of high-stakes case. Stekloff came out swinging, pointing out a minor error I had made in the Hardeman expert report. It was an inadvertent and inconsequential mistake, and I swatted that away. Then he moved on to the meaning of the term "differential diagnosis," which, as I've already mentioned, is a legal term and, as such, carries a meaning different from the one used by doctors. Lawyers use it to establish the cause of a disease, and, as I said earlier, I thought "differential etiology" was a better term for the legal process. Stekloff and I went back and forth discussing these terms until Chhabria interjected, and we had something I hadn't anticipated—a light, almost friendly exchange:

Judge Vince Chhabria: It sounds like the upshot is that the courts and the lawyers should not have started calling these "differential diagnosis," and it was a mistake on the part of the courts and lawyers; and we should stop doing this, and we should start calling it "differential etiology" because that's what it is.

Chadi Nabhan: I would never testify in this court what you should do. I wouldn't do that. I know better, but maybe you should have consulted us.

Judge Chhabria smiled at that. Was it possible that he was warming up to me a bit? No way, I thought.

Stekloff then moved on to a classic Monsanto play by asking me if there had been any evidence of glyphosate in Hardeman's body at the time of diagnosis. I answered that no, there had not been.

He followed up by asking me if I would therefore doubt causation related to glyphosate. In reply, I cited epidemiological studies that showed the association between glyphosate and non-Hodgkin lymphoma and stated that they did not measure serum levels of glyphosate; furthermore, many patients who were diagnosed with cancer had stopped using the compound several years prior to their diagnosis. Stekloff asked me to look at the deposition I'd given in November 2018 related to the woman who was diagnosed with lymphoma whom I mentioned earlier, and he suggested that I'd answered the same question differently then. I explained that my answer at the time implied that any patient who was still actively using the compound would have had measurable levels of glyphosate in their body had they been checked. Furthermore, I argued, that didn't contradict anything else I'd said:

Chadi Nabhan: *At some point through 20 years of being exposed to a particular compound, you must have had the offending material in the body. It doesn't have to be the day you were diagnosed or the year before you were diagnosed. That, we don't know. At some point it was present in the body, and most of the studies don't really look at that.*

Brian Stekloff: *You didn't give any of those explanations when you were asked about Ms. Gordon [the woman in the other case], right?*

CN: *Nobody asked me to explain exactly what it is, but that's exactly what is implied.*

BS: *Okay. All right.*

CN: *I mean, we give chemotherapy, the chemotherapy is in the blood, right? We don't always check the level of chemotherapy unless we need to, but it is present; and then it disappears from the body after you finish chemotherapy, but the possible toxicities that happened with chemotherapy years later occurred despite* *the fact that chemotherapy isn't in the blood. It is already gone. You were exposed to it. At some point throughout your medical journey or whatever you were doing, you would be exposed to these compounds and materials and if you test, you would find them. They just don't need to be tested the day of the diagnosis or the year before the diagnosis.*

BS: *Ms. Gordon also used Roundup for several years, right?*

CN: *She did.*

BS: *Then you were asked if at the time of her cancer diagnosis glyphosate was not in her body, would you—would you be able to say that glyphosate was a cause; and you said, "I would have serious doubts." That was your answer.*

CN: *I'm explaining to you what I meant by the answer, and I'm explaining to you that at some point you would find the actual compound in the body if you look for it.*

A Daubert hearing is all about methodology. Monsanto's goal was to show that my methodology was flawed, and my objective was to show Judge Chhabria that my methodology was sound. There were no jurors, just the judge, who listened intently and took notes as the cross-examination continued.

BS: *Dr. Nabhan, if a patient's exposure fits within the epidemiological literature that you have said links Roundup to non-Hodgkin lymphoma, and if there are no other risk factors, then you will not dismiss Roundup as a substantial contributing factor in a particular patient, right?*

CN: *You have to look at each case individually, but yes, you cannot dismiss it automatically and ignore a risk factor. It is like, you know, if you have somebody with lung cancer and is a smoker. You can't dismiss smoking. You have to put it on the list because you know smoking causes lung cancer. You can't dismiss it, no.*

I had made all these points before, but as Stekloff and I were going back and forth on when to include or exclude Roundup as a possible cause, Judge Chhabria interjected to ask a question. If no other risk factor existed in a hypothetical case, he asked, would I automatically conclude that Roundup had substantially contributed to the development of the patient's non-Hodgkin lymphoma?

"As a clinician, I would never make any determination automatically," I replied. "It is just not the way our clinician brain thinks. We have to be very analytical in each particular case. But if everything else is ruled out completely, it will be very hard to ignore the possibility that it was a contributing factor to that particular patient." Once again I tried to illustrate this by making a comparison to other diseases: "If a person develops a heart attack and you can't find any other risk factor and there is hypertension, you can't say that hypertension did not contribute to the development of heart disease."

All these convoluted sentences, all these double negatives: I was coming to understand something every first-year law student learns, which is that nuances matter, words matter. And words can be twisted almost beyond recognition.

To his credit, however, the judge was clearly trying to understand the issue.

VC: Let's say you rule in five potential risk factors and then you rule out four of them. You are left with one potential risk factor. Does it matter how strong a risk factor that is before you conclude that that risk factor caused the disease?

CN: Absolutely . . . you have to be convinced as a clinician that all of these factors have evidence in the literature that they should belong in that big basket that you are putting them in.

I thought that was a pretty straightforward answer, but deep down inside, I felt like the entire courtroom was waiting for a "gotcha," when the opposing counsel would announce, "See? His methodology is flawed. Get him out!"

The judge pressed on:

VC: *For example, you get lung cancer from a variety of different things . . . working in an asbestos mine, smoking, and then some lesser causes; and you might rule out all of those in at the front end when you are looking at a patient's history. Am I right about that?*

CN: *Sure. You look at the weight of the evidence and rule things out.*

VC: *So, if you rule out a bunch of stuff but there is one risk factor remaining, I assume that you have to ask before concluding that this is the factor that caused the person's cancer, you would have to ask how strong is the evidence that this exposure causes this particular cancer.*

CN: *Absolutely. You have to look at the general evidence about this particular risk factor, and you have to be convinced that the evidence is strong enough for you to keep it in despite the epidemiologic threshold that we talked about. I agree with that a hundred percent.*

VC: *Does that mean—coming back to the Roundup/NHL context— that whenever a patient has been exposed to Roundup above the threshold that you have identified and whenever you have ruled out all the other risk factors, that you would never conclude that we don't know what caused this person's NHL? You would always conclude that it was Roundup?*

CN: *So, it is my opinion that the evidence that links Roundup to non-Hodgkin lymphoma is strong to start with, and I obviously recognize that there are other folks in this courtroom that disagree with me; and that's why we are here. But it is my opinion that the evidence that links Roundup glyphosate to non-Hodgkin* *lymphoma is actually strong. As you know, Your Honor, I have looked at the literature and I have reviewed the literature and I have testified to the literature. My opinion is that the evidence is actually strong between non-Hodgkin lymphoma and Roundup.*

I again tried to make a comparison between glyphosate and NHL, on the one hand, and smoking and lung cancer, on the other. But Chhabria objected to this comparison:

VC: *But the evidence between the link of smoking and lung cancer—I assume you would agree—is much stronger than the evidence of the link between Roundup and NHL.*

CN: *I think you are correct; that the evidence is strong. But if we go back to thirty years ago, there were a lot of people who would have disagreed with you . . .*

I paused in mid-sentence. What flashed through my mind was this: "Did I just say people could have disagreed with Judge Chhabria? Have I lost my mind?" I took a breath and carried on.

CN: *Back then, doctors were actually smoking in the hospital in medical rounds. We have pictures of that. So that may be accurate today and—but if you had said that to somebody thirty years ago, they would have laughed at us. They would say there was no evidence of that. It is my opinion that the evidence actually is very strong between non-Hodgkin lymphoma and Roundup; and as a clinician, I cannot ignore that. And, again, I recognize that we may disagree on some of these things. I belong in the camp that says the evidence is strong, and I promise you thirty years ago, many of us did not think that smoking was linked to any cancer; and it took a lot until now—nobody even disputes that.*

After we'd finished discussing this issue, the matter of idiopathic non-Hodgkin lymphoma surfaced once more. I had addressed that in every single deposition I gave, and on the stand when testifying in the Johnson trial, but Monsanto brought it up again trying to cast doubt on my conclusion that the lymphoma suffered by Hardeman (and others) was caused by glyphosate. Here I reached for another comparison, explaining that when we were trying to determine what had caused a heart attack in a young patient, only after every known risk factor is ruled out would you tell the patient that their heart attack was of unknown cause.

Then Stekloff pivoted back to hepatitis C, and the questioning got heated. When he brought up a study to show the relationship between hepatitis C and non-Hodgkin lymphoma, he was unhappy that I started reading, directly from the paper, the limitations of the study as they were described by the authors. But I pushed back.

BS: *Dr. Nabhan, I understand that you believe there are limitations to this study. My questions are different, so will you please just try to answer my questions?*

CN: *I'm a clinician and a researcher. When you show me a paper and you ask me to comment on two lines in the paper, I need to put things in context. I mean, it's not fair to just show me conclusions and say "hepatitis C virus confers 20 to 30 percent increased risk" not taking into context the methodology, the other limitations of the paper. I mean, you can't just pick and choose the lines that you think they suit what you're trying to tell me and ask me to comment on them inappropriately. I mean, I have to acknowledge the limitations.*

Judge Chhabria had a few follow-up questions, the first of which centered on the latency of hepatitis C–associated lymphoma. I explained that latency varied widely, but I also suggested that oftentimes we see some degree of cirrhosis before we see lymphoma, and cirrhosis was something that

Hardeman did not conclusively have at the time of his lymphoma diagnosis.

At this point I felt I needed to clarify to the judge my position about hepatitis C and lymphoma once and for all. Hardeman had likely contracted hepatitis C in the late 1960s or early 1970s but had not been diagnosed with the disease until 2005, at which point he was effectively treated for it. He was diagnosed with lymphoma ten years later, in 2015.

"When you look at the details of this particular case, you have a situation of thirty-five years in which this hepatitis C caused no NHL," I said. "It did cause a mild degree of cirrhosis . . . and then was treated appropriately and adequately with sustained viral response for ten years."

I pointed out that all of Hardeman's testing showed his liver function was consistently normal and his hepatitis C had never reactivated despite receiving chemotherapy to treat his lymphoma. Thus, I concluded, hepatitis C was not the cause of Mr. Hardeman's lymphoma. "That may be different in another case, but in this particular case when you look at the particular situation, it would be very difficult for me to even understand how anyone could blame hepatitis C," I said.

"For the virus to be able to cause non-Hodgkin lymphoma, it has to be present somehow," I went on. "The reason we treat hepatitis C is because we believe the continuous presence of the virus is what causes the problem and what causes the damage to the liver. That is the essence of why we treat these viruses."

I thought this was clear, but Judge Chhabria did not seem convinced. This led to a long and technical explanation on my part about how viruses work, and how they can damage cells in a different way than a substance like glyphosate can.

Toward the end of my time on the stand, after both sides' attorneys had concluded their examination of me, Judge Chhabria

still had some questions for me. He clearly had done his homework and read enough studies to understand the science—more
than I had ever expected. He asked me to pull one of the studies
from the binder that Monsanto provided. The particular paper
he was referring to had investigated how hepatitis C was implicated in transforming indolent (non-aggressive) lymphomas into
the aggressive kind. Edwin Hardeman did not have an indolent
lymphoma that became an aggressive one; rather, he had an aggressive lymphoma from the get-go. I explained this nuance to
the judge, and then I went on to talk some more about how fast
this transformation can happen in patients with indolent disease
because I sensed that the judge was genuinely interested in trying to understand the mechanism behind this process. However,
I did not think that any of my answers satisfied him.

I was as frustrated as he probably was. Once again, I'd tried
my best to make my arguments as clearly as possible. And once
again, I seemed to have been unsuccessful in doing that with
Judge Chhabria, to the point where I wondered again why I'd
gotten myself involved in this.

All that said, I still would love to have a beer (or sparkling water, my favorite drink) with Judge Chhabria one day and talk science and law; I'm sure we'd have fun.

12

The Second Trial Begins

The Hardeman trial started on February 25, 2019, and in its own way it was as historic as the Johnson trial, as Hardeman's was the first federal case against Monsanto. Reporters from *Bloomberg*, the *Wall Street Journal*, the *Guardian*, and other news organizations were all present by 8 a.m. when the doors of the courtroom on the seventeenth floor of the Burton Building opened. And, as it turned out, the first day offered drama—but perhaps not the kind that many of us were expecting.

Plaintiff attorney Aimee Wagstaff started her opening statement by introducing the jurors to Edwin and Mary Hardeman and explaining how the couple had met one New Year's Eve. Right off the bat, Judge Chhabria didn't seem very happy with this warm and fuzzy tone, and asked Wagstaff to limit her opening comments to what pertained to causation, which was the focus of phase one of the trial; as we've seen, this part of the trial was supposed to be focusing on the general question of whether glyphosate can cause lymphoma in general and in Mr. Hardeman's case in particular.

Wagstaff moved on to explain to the jury that they would hear from three expert witnesses during this first phase of the trial: epidemiologist Beate Ritz, former CDC official Christopher Portier, and hematopathologist Dennis Weisenburger. She then showed a slide presentation that described the epidemiological literature. Judge Chhabria felt obligated to remind the jury that lawyer statements are not considered evidence.

Next, Wagstaff explained to the jury how the EPA and IARC had reached their differing conclusions about whether glyphosate posed a risk of cancer. At this point it was clear that Chhabria once again was unhappy with how Wagstaff was approaching these topics, especially as the slides that she was showing the jury had not been preapproved, as is customary. During the morning break, while the jury was not present in the courtroom, Chhabria lit into Wagstaff:

> *Ms. Wagstaff, you have crossed the line so many times in your opening statement, it's obvious that it's deliberate. The last time, the most recent time was when you were talking about the EPA, and you were referring to the EPA being vulnerable to political pressure. Totally inappropriate. Totally inappropriate. Totally inconsistent with everything we've discussed over the past several months. So, I'm going to give you one final warning. One final warning. If you cross the line one more time in your opening statement with respect to Phase I, if you bring in material during your opening statement that is inadmissible during Phase I, your opening statement will be over. I will tell you to sit down and I will tell you that your opening statement is over, and I will do it in front of the jury. Do you understand?*

Everyone could sense the heat in the courtroom as the trial resumed. As Wagstaff continued her opening statement, she tried to discuss Monsanto's involvement in the scientific studies and the communication Monsanto had with the EPA. Once more Chhabria was displeased. He felt that Wagstaff was challenging him and his instructions. Eventually, in what one news report would call "an extraordinary move," the judge threatened to sanction Wagstaff and fine her.[1] When Wagstaff remarked that she might have made Chhabria upset, he responded, "It's not about being upset. It's about running an orderly trial." While he ultimately allowed her to finish the opening statement, it was clear that what she had done stuck in his craw, as he ended the day by calling Wagstaff's behavior and statements "incredibly dumb."

In Stekloff's opening statement, he emphasized that non-Hodgkin lymphoma was a relatively common cancer and that the jury would be hearing from numerous experts that in many cases the causes of non-Hodgkin lymphoma were unknown. As he put it: "People unfortunately, like other cancers, are diagnosed with this type of cancer, non-Hodgkin lymphoma, every single day and we don't know why. Doctors don't know why. Oncologists, who are the doctors that treat cancer; pathologists, who are the doctors that diagnose cancer when they look at a tumor on a slide, they do not know what causes cancer." He went on to say that there were no pathology tests, imaging studies, or physical exam findings that could determine that glyphosate might have caused someone's lymphoma.

Stekloff appeared confident and articulate in his opening statement, which addressed the customary talking points that Monsanto always used. He emphasized that none of Hardeman's treating doctors had stated that his lymphoma was caused by Roundup; further, his treating oncologist had never noted in the medical records that Hardeman even used the product. He spent quite a bit of time discussing how hepatitis C was a major risk factor for lymphoma and told the jury that there were other confounding risk factors in Hardeman's case that might have played a role in the disease. He asserted that the plaintiff counsel wouldn't be able to convincingly show a link between Roundup and lymphoma in general and, more importantly, that they wouldn't even be able to show the link between Roundup and Mr. Hardeman's disease, given the presence of these confounding risk factors. He then told the jury that he would have two experts on the stand defending Monsanto's position, and he praised their academic accomplishments. Not surprisingly, Stekloff described the AHS as definitively ruling out an association between Roundup and lymphoma. My name came up, too—and he made a point of saying that while I was supervising the training of doctors doing a fellowship in oncology, I had never told any of them that Roundup caused lymphoma. He concluded by circling back to the point that none of the physicians treating Hardeman had ever brought up Roundup and claimed that this should be very powerful testimony that Roundup did not cause the disease in question.

Dr. Beate Ritz was the first expert called to the stand. I have become a fan of Ritz's because of her articulate way of describing the evidence, her level of knowledge, and the confidence she exudes. As Wagstaff introduced Ritz to the jurors, she noted that Ritz has a doctorate in epidemiology from UCLA and a medical degree from the University of Hamburg. She also made a point of highlighting that the governor of California had appointed Ritz to an expert panel tasked with assessing reports produced by the California Office of Environmental Health and Hazard Assessment, which works to keep California's air clean and free of toxins.

Ritz described to the jury the epidemiological literature on glyphosate and cancer, and she explained concepts such as statistical significance and confounders and how these apply to the Roundup/lymphoma research studies. She went out of her way to emphasize to the jurors that while statistical significance is an important criterion in biomedical research, as in all scientific research, it is not the sole thing that matters; furthermore, it cannot be taken out of clinical context.[2]

Wagstaff then asked Ritz to explain how studies dealt with the issue of the extent and timing of people's exposure to Roundup. This is a critical question, and I'll use Ritz's example of smoking to show how important this is. Let's say you ask someone the simple question "Have you ever smoked?" and they say yes. Ritz's point is that it's also essential to know how much and how long a person smoked. As she put it: "[If] you ask a smoker, 'Are you a smoker?' and he says yes, [then] that's it. He is a smoker. But you could also say, 'Well, how many cigarettes have you ever smoked?' And the answer could be, you know, 'When I went into the military, I tried it for a month, and, you know, it didn't become me and then I stopped.' But [even in that case the answer to] the question 'Have you ever smoked?' would have been yes. So, you classify somebody who . . . tried smoking for a month as a smoker. And then you have your neighbor who you have seen smoking every single day on the balcony." Her point was that not all exposures are created equal. In terms of exposure to Roundup, she explained the importance of knowing details such as how long the exposure lasted, how much was used, what protective attire was worn, and whether any spilling events happened. "All of that information then goes into how much I think that person actually got exposed," she concluded.

This smoking example was how she explained to the jury the concept of "dose response." The more you smoke, the higher your risk of smoking-related disease is, and the same applies to Roundup: the more you use it, the higher your risk of Roundup-related disease is. To most people, this sounds intuitive, but in a court of law everything has to be spelled out.

Wagstaff's next task was to walk Ritz through an explanation of how to assess confounders' impact on outcomes. As we've seen before, a "confounder" is another, unknown factor that affects a potential cause-and-effect relationship. It's always challenging to know whether jurors understand these nuances, but Ritz had the knack of explaining complicated ideas in an easily understandable manner.

At that point the jury was dismissed for the day, having been told that Ritz's testimony would continue the following morning. After they had left, Chhabria decided to address what he perceived as Wagstaff's misconduct. Jennifer Moore, who was another one of Hardeman's lawyers (and the co-lead counsel in this trial), stepped up to defend Wagstaff. When Moore asked what in particular Chhabria considered misconduct, Chhabria said that Wagstaff had brought up topics that were to be covered only in phase two of the trial; she was supposed to stick to general causation, but instead she had said quite a bit about Hardeman's personal history and his illness, which was supposed to be reserved for the later phase. Chhabria also said that Wagstaff had discussed the IARC data beyond the limits he had imposed, and that he was very unhappy with Wagstaff telling the jury that the EPA is subject to political pressure.

Then Chhabria's criticism of Wagstaff took a surprising and distinctly personal turn. He said that when he had initially criticized her for her conduct, she had not looked surprised at all. He believed that Wagstaff must have "braced herself for criticism" because she had intended to violate his pretrial orders. Furthermore, Chhabria referred to Wagstaff's "steely body language and expression." To some observers, this came across as criticizing her composure and, furthermore, taking that composure as evidence that she had deliberately intended to defy his pretrial orders. Wagstaff defended herself appropriately, telling the judge that she simply had remained composed, and she questioned why having good composure was being held against her. After some more back-and-forth between Chhabria and the plaintiff attorneys—all in the presence of Hardeman, whom the judge said was ultimately responsible for his lawyers' actions—the judge issued an order for sanction imposing a $500 fine, but with additional discussions to follow.

The next day, Ritz's testimony continued as Wagstaff resumed her direct examination. Wagstaff referred to how vital information was missing from the AHS, and asked Ritz to explain why it was missing. "And so, your criticisms about not knowing that information, how does that affect the data collected on the AHS?" she asked Ritz.

Ritz pointed out that in the study, the researchers "went through 'never-ever' use, right? 'Am I a glyphosate user?' Yes or no. And then we had, 'How many years did you use or how many days per year did you use?'" As she described it, the researchers had realized that because it was so difficult to determine whether someone had been exposed, they decided to ask questions that had more clear-cut answers, such as what protective equipment people had used and how they had applied the pesticide. "And then they came up with a very fancy method of saying, 'Okay. Your exposure intensity, how much you were exposed, depends on what you reported as protective equipment and how you applied.' But they applied these data, these pieces of information, to every pesticide they report, without knowing that for the one pesticide they did this and for the other pesticides they did that. They just applied it across the way because that's the only information they had." She went on: "So, somebody may have applied glyphosate, not worn protective equipment, applied it in a manner that exposed them maximally, but also applied another pesticide that the farmer considered much more toxic . . . and thought you should handle carefully. So, in that case, they are actually putting on a Tyvek suit and goggles and they are using a tractor. And when they are reporting, they say, 'Yes, I used a Tyvek suit and a tractor,' but that is for the other pesticide they use."

The researchers analyzed the combinations of different pesticides, different types of protective gear, and different methods of application using a statistical technique called intensity weighting. "That information then subtracts from glyphosate use. So, instead of saying this person was X number of years glyphosate exposed, it's intensity weighted, meaning we are subtracting 50 percent or 90 percent of exposure from that person" because of how they protected themselves against exposure to another type of pesticide.

She then went on to discuss how changes in exposure patterns over the years of the survey can lead to misclassification in "exposed" versus "non-exposed," especially as the use of glyphosate became more ubiquitous in 1996 several years after the survey was initiated.

> So, what is all what I'm telling you called? It is called exposure misclassification. And the best way to understand that is to think you have a white can of paint and a red can of paint; and every time that you get something wrong, you are mixing these paints. So, you are taking a cup and, you know, when you are getting it wrong, you are taking a cup of the red paint and you are putting it in the white. And guess what happens? You stir and it's a little pink. And then you take from the white can and put it in the red, and the red becomes a little lighter. If you do this often enough, we get pink in both cans. Meaning, we cannot distinguish the red from the white anymore or the exposed from the unexposed; right? And the more we have that situation happening, the less we can really say that that exposure caused anything.

Wagstaff asked Ritz about two papers in particular that had criticized the AHS's methodologies. One of these papers, on exposure misclassification, had been co-authored by John F. Acquavella. At the time of the paper's publication, Acquavella was employed by Monsanto as their top epidemiologist.[3] Another paper evaluating the AHS that was published in 2000 acknowledged the limitations of the AHS and suggested various improvements.[4] Both papers were critical of the AHS for many of the reasons I have already discussed. But Wagstaff was very strategic about how she presented these papers to the jury: she did so immediately after Ritz had explained the AHS and its shortcomings, in order to highlight that other researchers were also critical of the AHS, *including researchers associated with Monsanto itself.*[5] And Wagstaff pointed out to the jury that Ritz had used yet another study critical of the AHS to explain to her students the limitations of a particular type of study (prospective cohort studies). But somehow all these critiques and concerns were supposed to be ignored once the results of the AHS were shown to be favorable to Monsanto.

Tamarra Johnson, representing Monsanto, cross-examined Ritz. She started by reminding the jury that Ritz was not an oncologist, and she made sure that the jury heard that Ritz had not treated patients since 1989; presumably the idea was that learning that Ritz had not been directly involved in patient care in thirty years would make the jury think less of her. It made me think of how George Lombardi had spent valuable time during my cross-examination in the Johnson trial detailing how I no longer was seeing patients and instead focusing on administrative and research work. Tamarra Johnson also wanted to emphasize that Ritz had not examined Hardeman's records and that she was not testifying on his specific illness. Surprisingly, the cross-examination was short—probably, I thought, because of how impeccable Ritz's testimony was. There wasn't much that Monsanto could do to counter the points Ritz had made.

Before the jury left for the day, Chhabria reminded them of a number of stipulations—that is, facts and issues that both sides had already agreed on. Stipulations are designed to make the process more efficient, as the parties don't have to spend time arguing about facts that both had agreed upon prior to the trial.

In this case, he told the jury that both parties had stipulated that their experts have been paid a significant amount for their time in accordance with normal and customary rates. He further explained that the purpose of this stipulation was to avoid having to ask a bunch of questions of the expert witnesses about their compensation, so that the focus could be on substantive matters and the science.

The following day was scheduled for Dr. Portier's testimony. Unfortunately, several weeks before the trial, Portier had fallen ill while in Australia, and his physicians recommended that he not travel. Therefore, the plaintiff and defense teams had gone to Australia and spent a few days taking his testimony. The jury would hear his opinions via video.

Portier's testimony was centered on animal studies. On direct examination, he reviewed rat and mouse studies and explained eloquently how glyphosate caused lymphoma in rodents. He also reviewed the proposed mechanisms by which glyphosate can cause damage in the cells, leading eventually to cancer. Portier's video on his cross-examination was to be shown to the jury the following day.

However, the buzz on that day was not Portier's testimony but rather Judge Chhabria's exchange with Aimee Wagstaff, and the statement that he issued:

> *I was reflecting on the OSC hearing last night, and I wanted to clarify one thing. I gave a list of reasons why I thought your conduct was intentional, and one of those reasons was that you seemed to have prepared yourself in advance for—that you would get a hard time for violating the pretrial rulings. In explaining that, I used the word "steely," and I want to make clear what I meant by that. I was using "steely" as an adjective for steeling yourself, which is to make yourself ready for something difficult and unpleasant. My point was that I perceived no surprise on your part; and since lawyers typically seem*

surprised when they are accused of violating pretrial rulings, that was relevant to me on the issue of intent. But "steely" has another meaning as well, which is far more negative. And I want to assure you that that's not the meaning that I was using nor was I suggesting anything about your general character traits. So, I know you continue to disagree with my ruling and my findings about intent, but I wanted to make that point very clear.

Portier's cross-examination was rather uneventful and centered on Monsanto's attempt to show the jury how Portier opposed the EPA's views on glyphosate. There was nothing particularly new there; Monsanto had tried much of the same thing during the Johnson trial. However, the cross-examining of Portier led Judge Chhabria to allow the jury to watch an edited video of Portier discussing the report James Parry had prepared, at Monsanto's request, on the question of whether glyphosate was genotoxic. In that video, Portier explained that Parry had indeed found evidence of glyphosate genotoxicity, and that he had recommended that Monsanto do additional studies. We all know that none of these critical studies were done. The jury paid close attention to the video, though their facial expressions did not indicate what they thought of it. After Portier's testimony, the jury viewed Hardeman's treating physicians' testimonies on video.

That day Kathryn called me in the evening to let me know that they didn't plan on calling me during phase one of the trial. She said that Wagstaff felt pretty good about how the trial was going and preferred to save my testimony for phase two, when I would explain to the jury about Hardeman's condition, his treatment, and the effects of the chemotherapy he had undergone. Of course, if the plaintiffs lost the first phase of the trial, there wouldn't be any second phase.

Edwin Hardeman, the most important person in this trial, then took the stand to share his story with the world. As phase one of this bifurcated trial was designed to litigate general and specific causation in Hardeman's case, he was to testify about his lymphoma diagnosis and his use of Roundup. If the jury ruled in Hardeman's favor in phase one, as noted, phase two would begin, dedicated to damages and liabilities. After Hardeman took the stand, plaintiff attorney Jennifer Moore asked him about his use of Roundup—how often he'd used the chemical, how long he'd used it each time, what he'd been wearing when he applied the chemical, whether the substance had ever spilled on him, and what body parts were exposed. I thought Hardeman articulated his case nicely, and the jury listened attentively.

Once Moore had finished guiding Hardeman through his testimony, Monsanto was up. I suspect it's always challenging for the defense when the patient himself is on the stand in a case like this. On one hand, defense counsel risk alienating the jury if they don't show any empathy; at the same time, they want to challenge the plaintiff and try to poke holes in their story. What would Monsanto do? I wondered. Nothing, as it turned out— Monsanto decided not to cross-examine Hardeman.

In subsequent days the trial moved along, with various expert witnesses making their appearances and both sides arguing over what could be admitted. I am always surprised by how much energy is spent on deciding what can and can't be shared with the jury. The citizen in me says, "Share it all! Why hide anything? If we want a fair verdict, shouldn't the jurors be aware of every relevant detail?" But that's not how trials function, I learned. Prior to Dr. Dennis Weisenburger's testimony, for instance, Judge Chhabria spent a lot of time discussing what was going to be allowed into evidence. Monsanto didn't want Weisenburger discussing the IARC, but Chhabria did not agree. The trial wasn't a fight between the IARC and the EPA, he ruled (correctly, in my view).

Weisenburger, a professor at the City of Hope National Medical Center and a hematopathologist, had spent several decades of his life studying the impact of pesticides on the development of non-Hodgkin lymphoma and had published extensively in the field. Jennifer Moore conducted the direct examination and guided Weisenburger through a review of the epidemiological literature. Toward the end of the direct examination, they had this exchange:

Jennifer Moore: Well, the jury has heard from Monsanto's attorney in opening that, you know, most cases of NHL, the cause is listed as unknown. Why didn't you just say you don't know the cause like in these other cases of NHL here for Mr. Hardeman?
Dennis Weisenburger: Because we identified a cause.
JM: And that cause?
DW: The cause is Roundup. More likely than not it is Roundup.
JM: And Dr. Weisenburger, based on your forty years of investigating and researching the cause of non-Hodgkin lymphoma, your extensive literature review, your review of all the data, your own publications—I think there are over forty about the causes of non-Hodgkin lymphoma—your review of the medical records and your interview of Mr. Hardeman, please tell the jury your opinion, within a reasonable degree of medical certainty: what is the substantial factor in causing Mr. Hardeman's non-Hodgkin lymphoma?
DW: I think it is Roundup.
JW: Do you have any doubt as to your opinion that Roundup was a substantial factor in causing Mr. Hardeman's non-Hodgkin lymphoma?
DW: No.

It continued to amaze me that none of Monsanto's employees or executives came in person to sit on the stand, face the jury, and defend the company. As in the Lee Johnson trial, they could have chosen to be in court in person (though they were not obligated to do so). An edited video deposition of Monsanto scientist Donna Farmer, arguably one of the biggest defenders of glyphosate, was shown to the jury. Though Dr. Farmer had been deposed several times, and for hours each time, the jurors saw only a few minutes of her deposition. Basically, they saw what the judge allowed them to see—including something that was a bit of a shocker: an email from 1999 showing that Farmer was critical of the AHS—the study that Monsanto's attorneys had held up throughout the Johnson and Hardeman trials as the gold standard of research (see figure 22). Monsanto had always made it sound like the AHS—which, as I've noted several times, found no association between glyphosate exposure and lymphoma—was essentially the greatest work of scientific investigation since the dawn of time. But back in 1999 Farmer had had a different opinion. "Small in scope . . . unreliable," she'd written in regard to the AHS. She even described how others had criticized the AHS as "junk science."

When Wisner asked her about that email, Farmer said that her mind had changed. You bet it had.

The plaintiff side rested, and now it was Monsanto's turn.

Dr. Lorelei Mucci, an epidemiologist from Harvard, tried to poke holes in every epidemiological study the plaintiff's expert witnesses had mentioned—except the AHS. When discussing the AHS, Mucci sang its praises and told the jury how the first author of the AHS had won an award for an outstanding research paper written by a staff scientist or clinician. Mucci emphasized that the 37–38 percent dropout rate (those who had been among the initial group but who had not filled out the second questionnaire) had no impact on the outcomes of the study, as statistical methods allowed for proper prediction of what they would have answered.

```
_____ Reply Separator _____
Subject: Re[4]: Questions about Glyphosate
Author:  DONNA R FARMER at MONSL125
Date:    5/31/99 1:55 PM


Tom,

Your welcome. Life is always busy....work/home/work/home...the key is the
balance!!!!

Regarding business....unfortuantely we feel that Hardell is just the
tip of the iceberg for these type of "association epi" studies.  We
have his two papers with NHL and hairy cell leukemia and one from a
Canadian Ag Health study that declares an association between
glyphosate and miscarriages and pre-term deliveries.

What is of greater concern however is an American initiative called
the AHS.

The AHS stands for Agricultural Health Study - a large multi-faceted
epidemiologic study being conducted by scientists with the National
Cancer Institute (NCI), the EPA, The National Institute for
Environmental Health Sciences (NIEHS).  It is its 7th year of data
collection and soon will publish results linking specific pesticides
to various health effects.  These organizations believe that farmers
and their families are suffering from a variety of illnesses and
that these illinesses are caused by pesticides....no bias there!

The widespread and ever growing use of glyphosate caused the AHS
investigators to reevaluate and give more priority to glyphosate.

It is a prospective study of 90,000 farmers and their families in
Iowa and North Carolina.  The primary purpose of the study is to
look for associations between pesticides and human health effects.

Many groups have been highly critical of the study as being a flawed
study, in fact some have gone so far as to call it junk science. It
is small in scope and the retrospective questioneer on pesticide
usage and self reported diagnoses also from the questioneer is
thought to be unreliable...but the bottom line is scary...there will
be associations identified between glyphosate use and some health
effects just because of the way this study is designed.
```

Figure 22: Portion of the email written by Donna Farmer in which she critiques the AHS in 1999.

When Aimee Wagstaff cross-examined Mucci, she noted for the jury that Mucci had been involved in a 2008 congressional investigation based on her assertion that the toxic chemical acrylamide was not risky; she also showed the jurors that two congressmen had expressed their concerns regarding Mucci's testimony about acrylamide, since at the time of her testimony there was already evidence refuting what she said. The exchange over this subject lasted a while, and Aimee conducted the cross-examination expertly, casting doubt on Mucci's testimony. In court, it's all about creating doubts in the minds of the jury regarding the opposing experts, and Aimee did just that.

The biggest win that Wagstaff secured against Mucci was when she brought up a textbook on cancer epidemiology that

Mucci had co-edited in 2018.[6] In one of the chapters, there was a list of carcinogens classified by IARC as group 1, carcinogenic to humans, despite limited evidence in human studies but strong evidence from mechanistic data. Wagstaff wanted to highlight that Mucci, in her own book, had acknowledged that some hazardous compounds can be carcinogenic despite limited human data.

March 11, 2019, was an important day in the Hardeman phase one trial, as Dr. Levine, an oncologist testifying on behalf of Monsanto, appeared in court. Dr. Levine is considered an expert in lymphoma and has done much work on HIV-related malignancies and has also held various administrative posts at the City of Hope National Medical Center including chief medical officer. As to be expected, the defense counsel bragged about Levine's accomplishments, awards, and publications. I thought it was interesting that Levine admitted that she did not have in-depth knowledge about pesticides outside of what she'd learned from reading the report Dr. Mucci had prepared for this trial as an expert witness.

One of the ways experts solidify their credibility with jurors is by criticizing the opposing experts' theories. To that end, Levine made sure jurors heard why she did not agree with Weisenburger's assessment that Roundup caused lymphoma. Levine's central argument was that Hardeman had had hepatitis C for many years, and that even though it had been effectively treated, it still must have been the most important cause of his non-Hodgkin lymphoma. Levine proposed that the many years during which the virus had been present in his body—he had been infected perhaps even as early as the late 1960s—had produced the cellular damage that eventually caused the lymphoma, and that his 2005 treatment for the virus had eradicated the virus but not repaired the damage to his cells. She framed this as a "hit-and-run" theory, meaning that even though the hepatitis C was treated, it had already "hit" the DNA, causing the damage, and then "run away," meaning the virus disappeared with therapy, but the effects of the hit were sustained.

Although Levine was convincing, coming across as articulate and authoritative, Jennifer Moore was on top of her game as she cross-examined Levine about pesticides:

Jennifer Moore: *Your specialty has not been lymphoma that's caused by pesticides; is that fair?*

Alexandra Levine: *That is not my specialty.*

JM: *When you have a patient come into your office that has non-Hodgkin lymphoma, do you ask them about their pesticide use?*

AL: *I do not.*

JM: *Is it possible that all those cases that you referred to earlier as being idiopathic or having an unknown cause, that it actually could be from pesticide use, but you don't know because you don't ask?*

AL: *That's a total assumption. I can't answer that question. I don't know.*

JM: *But without asking your patient about their pesticide use, aren't you assuming that the cause is unknown?*

AL: *No.*

Moore moved on to establish that Levine had looked at all risk factors published by the American Cancer Society as plausible risks for non-Hodgkin lymphoma *except* pesticides.

At some point in the testimony, Moore was reviewing with Levine how Weisenburger testified and reached the conclusions that Roundup caused Hardeman's lymphoma based on his exposure.

JM: *He had Roundup, which is a pesticide; right?*
AL: *I thought it was an herbicide.*
JM: *Do you know that herbicides are a type of pesticide?*
AL: *No.*

I could imagine, when I heard that, how uncomfortable it must have been for Dr. Levine to be on the stand and saying what she said.

While I was, of course, on Edwin Hardeman's side, I couldn't help but feel some empathy for a fellow expert witness. I knew how hard it was to sit there and be attacked. But I wondered how all of this would play with the jury.

13

Verdicts

The time had come for closing arguments in phase one of the Hardeman trial. Aimee Wagstaff, as counsel for the plaintiff, went first. She explained that she did not have to prove that Roundup was the *only* potential cause of Hardeman's lymphoma; all she had to show was that it was a *substantial contributing factor* to him developing the disease—a key distinction, I thought. She conceded that it was possible that his lymphoma had various causative factors, a point that would be critical when the jury started deliberating. She seemed to be giving them a way out by telling them that it would be possible for *both* Roundup and hepatitis C to have contributed to the development of Hardeman's lymphoma. It was a shrewd strategy, and a point that most medical experts would concede.

When Stekloff's turn came, he highlighted the weaknesses in the plaintiff's arguments, and emphasized that Hardeman's lymphoma was likely caused by hepatitis C. He cited Alexandra Levine's "hit-and-run" argument that the virus had already caused cellular damage, and treating it would only have eliminated the virus, not repaired the damage it had produced, which was a cause of the disease. Again, taking a page from the Monsanto playbook, he reminded the jurors that none of Hardeman's treating doctors had ever mentioned Roundup as a causative factor in their medical records. Stekloff also made sure that the jurors knew that the EPA had never classified Roundup or glyphosate as carcinogenic.

Now all we could do was wait for the jury to return a verdict. In Chicago, I found my anxiety increasing by the hour. The longer it takes for the jury to come back, the more speculation there is about what might be preventing the jury from coming to a quick decision. I learned that the plaintiffs usually are nervous because they start assuming that their case was shaky, and therefore the jurors are taking their time. And defense counsel are usually delighted as more time passes, as they might be dealing with a hung jury.

The jury started deliberating on March 13. Interestingly, during their deliberations they requested a readback of parts of the testimony, though what that signified was anyone's guess.

Around 2 p.m. on March 19, Judge Chhabria notified everyone that the jury had reached a verdict. The jurors entered the courtroom, and the verdict was handed to the judge. Chhabria read aloud so that everyone could hear. "Did Mr. Hardeman prove by preponderance of the evidence that his exposure to Roundup was a substantial factor in causing his non-Hodgkin lymphoma?" The jury's answer: yes (see figure 23).

The jury had delivered a major blow to Monsanto. And the win in phase one was even more powerful considering that Chhabria did not allow discussing Monsanto's alleged influence on research and regulations.

Judge Chhabria then announced that the court would be moving to the next part of the trial. "Phase two will be the final phase in the trial," he said to the jury, "and the issues that you will be

Figure 23: Edwin Hardeman after winning phase one of the trial. Aimee Wagstaff stands behind him at the press conference. Source: Reuters/Alamy Stock Photo

considering are whether Monsanto is legally liable for the harm caused to Mr. Hardeman and, if so, what the damages should be."

Chhabria then made a statement that I found quite impressive—and which seemed to be a departure from what his stance had appeared to be before:

> *Although the evidence that Roundup causes cancer is quite equivocal, there is strong evidence from which a jury could conclude that Monsanto does not particularly care whether its product is in fact giving people cancer, focusing instead on manipulating public opinion and undermining anyone who raises genuine and legitimate concerns about the issue.*

"He just called them a heartless corporation," I thought. "That can't be good for Monsanto." I expected that the Bayer executives in Germany needed more aspirin after hearing this. And no doubt Bayer stockholders were fuming.

Monsanto issued a statement after the phase one verdict: "We are disappointed with the jury's initial decision, but we continue to believe firmly that the science confirms glyphosate-based herbicides do not cause cancer. We are confident the evidence in phase two will show that Monsanto's conduct has been appropriate, and the company should not be liable for Mr. Hardeman's cancer."[1]

Phase two was like a completely new trial. There were opening statements, just like the first phase. In hers, Aimee Wagstaff reminded the jury why they were there. She emphasized what Monsanto had known about Roundup and that the company had failed to warn its customers of the known toxicities. She was able to remind them how Monsanto allegedly manipulated the science and allegedly influenced the EPA. And she pointed out that Dr. Donna Farmer, a Monsanto toxicologist, had in 2003 said in an email in response to an inquiry from an international colleague, "You cannot say that Roundup is not a carcinogen; we have not done the necessary testing on the formulation to make that statement. The testing on the formulations are not anywhere near the level of the active ingredient. We can make that statement about glyphosate and infer that there is no reason to believe that Roundup would cause cancer."[2] Wagstaff reminded the jury of Dr. Parry's recommendations and how Monsanto refused to perform any of the studies that he had suggested; Monsanto had already stipulated that they had not conducted any long-term animal carcinogenicity studies since 1991. She elaborated on how Monsanto had allegedly ghostwritten several studies and how its leadership allegedly endorsed this strategy.

For Monsanto's opening statement in phase two, Stekloff started off by accusing the plaintiff attorneys of cherry-picking data that aligned with their theory. He asserted that Monsanto had acted responsibly, and pointed out once again that the EPA, along with other regulators around the world, had determined that Roundup was safe. He told the jury of all the toxicologists, chemists, and epidemiologists that the EPA employed, all of whom knew what they were doing. The Monsanto strategy was very clear: whatever the EPA said must be accurate, as if the agency had never gotten anything wrong.

The jurors were shown various video depositions of Monsanto's employees and executives. The most notable one was that of Donna Farmer, who had been deposed by Brent Wisner. Wisner called attention to Monsanto's "Freedom to Operate" program, which dedicated resources to counter any emerging evidence against Roundup. But Farmer characterized the program differently: "Freedom to Operate is a program that allows people to buy our products." And when it came to the epidemiological studies that showed an association between glyphosate and lymphoma, Farmer's answers were not convincing and lacked confidence.

Wisner presented Farmer with an email that she'd written in response to a colleague who shared with her the results of Mikael Eriksson's 2008 study showing that exposure to glyphosate for more than ten days over the course of a lifetime more than doubled the risk of non-Hodgkin lymphoma. (See figure 24.) Farmer's response to her colleague clearly highlighted that Monsanto had this study on its radar and was anticipating that activists would use it against Roundup. She said, "We have been aware of this paper for a while and knew it would only be a matter of time before activists pick it up. I have some epi experts reviewing it. As soon as I have that review we will put together a backgrounder to use in response."

When Wisner started questioning Farmer about Monsanto's ghostwriting a scientific article, the exchange heated up.

Brent Wisner: *Is a form of ghostwriting somebody else writing a portion of it and not disclosing their involvement?*

Donna Farmer: No.

BW: You don't consider that ghostwriting?

DF: No, I gave you my definition.

BW: Would you call that deceptive authorship?

DF: I'd call it editing.

BW: So, if I prepared a paper at school and someone literally wrote paragraphs of that paper and I submitted that paper under my name, even though another person wrote portions of it; you would agree that is unethical, right?

DF: I am saying that I think you have to look at the contribution on a case-by-case basis.

Wisner used a couple of articles as examples to illustrate how Monsanto, and Farmer in particular, had engaged in ghostwriting. In fact, he showed her an email from the lead author of one of the papers in which the author stated, "Donna, you have added significant text to the document with regard to the following

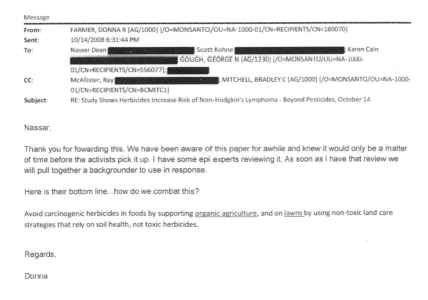

Figure 24: Donna Farmer's email about the Eriksson study.

references. Unless someone from Monsanto plans to be listed as an author, we need to see these references in order to verify that we are in agreement with the newly added text."

Kathryn Forgie had informed me that I would be testifying in phase two on March 22, so on March 21 I headed to the airport. The weather was ugly outside, as a major storm was brewing. I arrived well ahead of my 9 a.m. scheduled departure, but soon I saw that my flight had been delayed due to the impending snowstorm. As I stared out of the window, my anxiety rose as I noted that departure times on the flight board were being replaced by cancellations; three outbound flights to San Francisco had already been cancelled, and many incoming flights into O'Hare were unable to land. I phoned Kathryn and explained the situation; we even discussed options if I could not make it on time for the trial testimony the next morning. For a moment the thought of driving to San Francisco crossed my mind, but it evaporated as soon as I learned that it would take me more than thirty hours.

I headed to the customer service desk in the United lounge, and there I saw a familiar face: the widow of a patient I had cared for several years earlier. I hadn't known she worked for the airline. We talked a little bit about her late husband; she expressed so much gratitude about the care I had provided that I found myself blushing. I am always struck by how appreciative cancer patients and their families are to their treating physicians, nurses, technicians, aides, and everyone within health care who touches their lives.

Then I explained that I absolutely had to be in San Francisco the following morning for a court appearance.

"Are you in trouble, Doc?" she said with a mischievous smile.

"No, I'm expected in court, to testify as an expert in a patient case."

She couldn't change the weather, but she was able to find me a seat on an afternoon flight that would get me to San Francisco a little after 5 p.m.

After a rough ride, in which the flight attendants had been ordered to remain seated for the first thirty minutes (never a good sign), we made it to San Francisco. I jumped into an Uber and headed to the Airbnb that the attorneys had rented.

There, Aimee Wagstaff walked me through her proposed direct examination for my appearance the next day. The plan was to explain to the jury the medical course that Hardeman went through, the type of chemotherapy he received, and its side effects. "Remember, Chadi," she said, "your role here is not to convince the jury that Roundup was carcinogenic. That was already established during phase one."

I've already mentioned that during the trial Monsanto had stipulated that they had not conducted a long-term animal carcinogenicity study since 1991. During our preparations, Jennifer Moore shared with me another eye-popping insight from phase one: Monsanto had admitted that they had never studied the association between their glyphosate formulation and non-Hodgkin lymphoma. Both of these stipulations were to be read in front of the jury the next morning ahead of my testimony.

On March 22 I arrived at the Burton Building around 8:15 and met Kathryn in the lobby.

"Good morning," she said. "Ready for your close-up?"

I managed a weak smile. "As ready as I'll ever be. You know me—I just don't want to let Mr. Hardeman down."

She nodded and asked me to wait outside, as there were various procedural issues that needed to be addressed prior to my testimony. The jury also had to finish watching the video deposition of Donna Farmer.

It was close to eleven o'clock when Kathryn came outside and saw me pacing nervously back and forth in the corridor right outside the courtroom. It was time to testify.

I walked to the stand and was sworn in. My eyes met Chhabria's, and I nodded. I thought I saw a faint smile on his face—or maybe it was my imagination.

Wagstaff conducted the direct examination. She started with my least favorite part—asking me to tell the jury about myself. No matter how often the lawyers reminded me that this portion of the testimony was important in order to establish my credibility, I was extremely uncomfortable bragging about my accomplishments. Seriously, what accomplishments? Wagstaff had to do part of that job for me, circling back to remind the jury of my proficiency in lymphoma and my publications to make sure they knew that I was well qualified to discuss Mr. Hardeman's case.

She then asked me to explain to the jury what lymphoma was and to discuss the type of lymphoma that Hardeman had. How is the disease usually treated? And what type of chemotherapy did Hardeman receive? I answered in detail, describing each individual chemo drug that Hardeman was given, as well as the side effects that he experienced. One of them, as I've previously mentioned, was the bone pain from the growth factor injections that help keep white blood cell counts high enough to ward off infections. We also discussed the nausea and vomiting that resulted from the chemo drugs, which was something that jurors were more likely to have heard of. I went on to describe some of the possible long-term effects of chemotherapy drugs, such as secondary leukemia, other cancers, and possible cardiac issues.

Wagstaff asked me about Hardeman's prognosis, and I was happy to say that it was favorable. However, I explained the importance of making sure that all patients diagnosed with this disease received long-term follow-up.

Wagstaff wrapped up before 1 p.m. with the standard question asked of all medical experts: "Are the opinions you gave today given to a reasonable degree of medical certainty?"

"Yes," I replied.

It was now time for Monsanto's cross-examination. I took a deep breath, readying myself.

Then I heard Monsanto's counsel say, "We have no questions."

Judge Chhabria said, "Okay, great," and then sent us out for our lunch break.

I was stunned. Why had Monsanto passed on the opportunity to cross-examine me in front of a jury? It must have been a strategic choice, yet to this day I remain puzzled about what was behind that decision. Maybe it was because my testimony was focused only on the chemotherapy that Hardeman received, and there was nothing controversial in that. In any case, clearly Monsanto believed they stood to gain more by not questioning me.

I must give Kathryn Forgie credit here, as she'd predicted the night before that Monsanto would not cross-examine me. But of course, I had dismissed the idea.

Back in Chicago the following morning, I read press reports on Hardeman himself testifying in court right after I left. He corroborated the toxicities and side effects he'd suffered from the chemotherapy, commenting on hair loss, bone pain, and gastrointestinal side effects. Apparently, he was a bit emotional as he recounted his experience. Despite that, Monsanto decided to cross-examine Hardeman . . . and what they questioned him about was how frequently he had checked the Roundup label. I can't imagine that went over well with the jury.

On Wednesday, March 27, at 2:52 p.m. local time, it was announced that the jury had reached a verdict in phase two of the

Figure 25: Edwin Hardeman at his press conference after winning the $80 million verdict. To his immediate right is Aimee Wagstaff. Jennifer Moore is at the far left of the photo. Source: Reuters/Alamy Stock Photo

trial. Everyone gathered in the courtroom as Chhabria read the verdict:

Question 1: *Did Mr. Hardeman prove by a preponderance of the evidence his claim that Roundup's design was defective?*

Answer: Yes.

Question 2: *Did Mr. Hardeman prove by a preponderance of the evidence his claim that Roundup lacks sufficient warnings of the risk of NHL?* **Answer: Yes.**

Question 3: *Did Mr. Hardeman prove by a preponderance of the evidence his claim that Monsanto was negligent by not using reasonable care to warn about Roundup's NHL risk?*

Answer: Yes.

Then Chhabria read out the compensatory damages: past economic loss, $200,967.10; past non-economic loss, $3,066,677;

future economic loss, $2 million. On top of that, the jury decided that he was entitled to punitive damages and awarded him an additional $75 million.

"Wow!" was all I could think. It was a historic verdict. I felt a sense of pride that I had played a very small part in delivering a just verdict to Edwin Hardeman, just as I had with Lee Johnson. And I felt a sense of righteous satisfaction that Monsanto's tactics and seeming disinterest in the health of those who used its products had been exposed for the world to see. Hardeman and his lawyers were joyous and grateful in the press conference that immediately followed (see figure 25).

Monsanto/Bayer issued this official statement shortly after the verdict: "We are disappointed with the jury's decision, but this verdict does not change the weight of over four decades of extensive science and the conclusions of regulators worldwide that support the safety of our glyphosate-based herbicides and that they are not carcinogenic."[3]

14

The Pilliods

It was bright but brutally cold in Illinois on the morning of Monday, December 17, 2018. The snow that had fallen over the weekend had become ice; walking and driving were treacherous. I was in my office at Cardinal Health in Waukegan, awaiting the arrival of Alva and Alberta Pilliod from California, an older couple I would help defend in court in a few months.

In the spring of 2015 Alberta Pilliod had gone to the doctor because she was experiencing vertigo and had begun to be unstable when she was walking. This led to various tests that eventually confirmed diffuse large B-cell lymphoma, the aggressive form of non-Hodgkin lymphoma, but in her case, the disease was found only in her brain and nowhere else in the body; when this occurs, it's called primary central nervous system lymphoma (PCNSL). She was treated with aggressive chemotherapy over the next few months, and her disease went into remission in the fall of 2015. Less than a year later, however, her lymphoma came back, and another round of chemo was needed to put it back into remission. Her doctor had decided to keep her on an oral chemotherapeutic agent in order to prevent relapse, and it seemed to be working; when we met, she was still in remission.

What made Alberta's story unusual was that Alva, her husband, had the same type of diffuse large B-cell lymphoma, though his was not localized to the brain; it was in other areas of the body such as the lymph nodes and bones and was therefore classified as advanced-stage lymphoma. He'd developed it in 2011, a few years before his wife, at age sixty-nine. Alva too had been treated with chemotherapy, which put him in remission. Although not unheard of, it's quite rare for two individuals in the same household to develop the same type of lymphoma within a few years of each other. I'd seen hundreds of lymphoma cases over the years and could not recall another scenario where spouses had the same lymphoma.

The other thing Alva and Alberta had in common was that both of them had used Roundup extensively since 1982. No wonder the Miller Firm was eager to litigate this case against Monsanto.

I walked down from my office to greet them as they exited the taxi in front of the building. They looked a bit older and frailer than I had expected. Would they be able to endure a major trial and grueling questioning from Monsanto?

"Watch your step! Be careful!" Alberta warned her husband as they gingerly stepped out of the cab onto the icy sidewalk.

"I'm sorry we couldn't have arranged for better weather," I said with a smile.

"It's pretty chilly," said Alva. "But you know, it can get pretty nippy in the Bay Area this time of year, too."

"It's the snow and ice that we don't get out there!" Alberta chipped in.

I liked how they practically finished each other's sentences. Alva and Alberta had been married for more than thirty years.

After a bit of chitchat, we got down to the business at hand. I asked about their medical history, and they ticked off the details—dates, doctors, diagnoses. I knew of the extensive use of Roundup that they both had, but I wanted to capture more granular details about their use of the product.

"Oh, we were heavily exposed," said Alberta. "We used it a lot."

"Can you be a little more specific about that?" I asked. "How many days a week did you use it? Where were you spraying—and how many hours a day? Did you wear protective gear?" Those were the sorts of details that, I now knew, would be needed in order to get the facts and present a case that would hold up in the courtroom.

"I can get you this information, though I will need to dig a bit deeper in my memory," she said. "But please remember that we never thought that this was an unsafe product. There were no warning labels, and we had no clue that we might be using something that could cause cancer."

"I understand," I responded empathetically. "Do what you can, but these details are important."

I spent almost two hours with them, and as I finally waved goodbye and put them in a cab that would take them to the airport, I knew that this case would be difficult because of their age and the fact that they both had other medical problems. Alberta Pilliod had had superficial bladder cancer; the cancerous cells were excised, after which she received infusions of an immunotherapy drug called BCG into her bladder, and the disease had never come back. And at the time of her lymphoma diagnosis, her BMI had been 38, which is in the obese category. Alva had had several occurrences of skin cancers that usually result from excessive sun exposure (early-stage melanoma and a few squamous cell cancers). He also had a history of meningitis that had led him to develop seizures at a young age. Furthermore, Alva had a mild case of ulcerative colitis, which is a disease that causes inflammation of the bowel; it's thought to be autoimmune, meaning that it might be caused by the body mistakenly identifying its own cells as abnormal and attacking them. Patients with this disease sometimes are treated with drugs that suppress the immune system, but Alva had never received such powerful drugs. I was fully expecting Monsanto to claim that these other medical issues were the cause of their lymphoma, not Roundup. It was going to be another wild ride.

My deposition in the Pilliods' case was scheduled for January 2019, shortly after I had returned from Syria, where I'd spent ten days visiting family. At the time, I was also considering a new employment opportunity: joining a consulting firm that focused on gathering real-world insights and seeing how these apply to oncology care. I have always been interested in real-world patterns of care, which oftentimes might not be similar to what happens in clinical trials. Most patients with cancer are not treated in clinical trials, but rather in routine care, so joining this firm and leading its clinical operations and strategy fit my growing academic interest. Exploring a new opportunity is usually exciting, but my mind was beyond occupied with this litigation and I was running out of hours in the day to manage all my obligations.

I met Curtis Hoke, an attorney from the Miller Firm who was to represent me as I was being deposed in the Pilliods' case on January 19, 2019, in the lobby of the Marriott Suites in Deerfield, Illinois, which was close to my house. Curtis and I chatted briefly before we walked together into one of the meeting rooms on the first floor of the hotel, where I met the attorneys representing Monsanto: Joe Tomaselli, whose pleasant demeanor was a welcome departure from my prior experiences with Monsanto attorneys, and his colleague, whose business card indicated that she was both an attorney and an MD. I knew her medical background would be a clear asset to Monsanto during these deposition proceedings.

"Wow," I said to her. "Impressive. You know, I wanted to go to law school, but time wouldn't allow it. I had to resort to watching Netflix legal dramas instead." That got me a pale smile in response, but nothing else. Okay, I thought. Let the chess match begin.

Tomaselli started by asking general questions that focused on my methodology and how I had come to conclude that glyphosate caused non-Hodgkin lymphoma. These were the same questions every Monsanto attorney had asked at previous depositions; I suspect they were hoping to hear my answers change, to provide something they could jump on.

Tomaselli then asked me if I knew what Alberta Pilliod did for a living before she retired.

"Of course," I said. "She worked as a school administrator."

Tomaselli smiled. I got the distinct impression he was ready to declare checkmate very early in the match.

"Did you know about the relationship between schoolteachers and the development of non-Hodgkin lymphoma?" Tomaselli then moved to admit an exhibit into evidence: a 2007 paper that suggested that there was an association between employment as a teacher and increased risk of non-Hodgkin lymphoma.

"I haven't seen this before," I said. "But I would question any paper that would propose such a strange observation." The premise and conclusion seemed almost preposterous to me. Still, in the back of my mind I wondered if I'd missed something, despite all those hours I'd spent poring over the research.

I took the time I needed to review the paper and gather my thoughts. Now it was my turn to move the pieces on the board. I asked Tomaselli if, based on this paper, schoolteachers had ever signed a consent form that acknowledged their lymphoma risk when they signed up to teach at any school. Of course, I knew that they hadn't. I then read aloud, for the record, what the authors stated about the limitations of their investigation: "For no job or industry is there, at present, conclusive evidence of a causal association." In other words, there was no relationship between lymphoma and being a teacher—or most other white-collar occupations, including being a lawyer or a judge.[1]

It's important to note that the paper Tomaselli showed me was a meta-analysis, which is a statistical analysis that looks at the results of multiple previously published studies asking the same question. The idea is that while any given study might not have been able to conclusively answer the question, by lumping them together maybe we can. And, when it came to schoolteachers and lymphomas, the authors of the meta-analysis did say that the observed association could be ascribed to methodological problems, not to some causal link.

Meta-analyses can be a very useful tool, and there were other such studies that supported a link between glyphosate and non-Hodgkin lymphoma. One looked at research articles published between 1980 and 2014 showing an association between glyphosate and B-cell lymphoma.[2] Another study—this one sponsored by Monsanto, the manufacturer of Roundup—explicitly stated in its abstract: "This systematic review and meta-analysis rigorously examines the relationship between glyphosate exposure and risk of lymphohematopoietic cancer (LHC) including NHL, Hodgkin lymphoma (HL), multiple myeloma (MM), and leukemia." And in their discussion and in their conclusion the authors stated that they found "borderline" and "marginally" significant associations between glyphosate and NHL. That this was the conclusion in a study *sponsored by Monsanto* cannot be ignored or downplayed.[3]

Next Tomaselli moved on to the Pilliods' age. I sighed and repeated what I'd said about age in prior depositions. "Age is a risk factor for any and every disease under the sun," I said.

But Tomaselli wasn't satisfied with that answer. He asked a torturously worded follow-up question.

Joe Tomaselli: *Simply, a person presenting with [non-Hodgkin lymphoma] at or around the time of being seventy years old can fully explain their presentation?*

Chadi Nabhan: *If you're asking could somebody at the age of seventy present with (NHL) of unknown cause? Yes. We can't ignore another known cause and just blame it on your age. That's not fair to older people.*

JT: *Would you agree that Mr. Pilliod could have developed the exact same DLBCL at exactly the same time with exactly the same features and had exactly the same course of treatment and recovery even if he never used Roundup?*

Curtis Hoke: *Objection to form.*

CN: *I stopped playing God a long time ago. I think any of us could develop any type of disease anytime, including Mr. Pilliod. So the answer is yes, I mean, anybody could develop anything.*

JT: *The exact same cancer and course that Mr. Pilliod had exists in people who never used Roundup?*

CN: *Correct.*

I was growing tired of Monsanto wanting to blame age for lymphomas seen in any older patients. But I knew this issue would be brought up during the actual trial and I should be ready with an effective counterpunch.

Tomaselli then shifted to Alva. Monsanto wanted to propose that he was somehow immunocompromised. Couldn't that have been a cause for his lymphoma? And what about the skin cancers that Alva had had? As Tomaselli tried these new lines of questioning, showing some papers that associated prior skin cancers with lymphomas in some patients, I noticed the physician-lawyer scribbling notes, which Tomaselli occasionally glanced at. While silent, the doctor was clearly playing a role in this deposition.

It irritated me that I felt Monsanto was trying to link everything under the sun—except glyphosate—to lymphoma. So, I felt this was an appropriate time for me to point out that there is a big difference between association and causation. Just because two diseases are associated with each other, it doesn't mean that one causes the other. I rolled out my favorite example. "Cataracts, for example, is a disease of older patients, and certainly we all know that pancreatic cancers also occur in older patients," I said. "But does this mean that cataracts cause pancreatic cancer? Of course not."

By this point, we'd reached the three and half hours that Hoke, the Miller Firm lawyer, had insisted was the maximum allowable time for my deposition. The Monsanto lawyers disagreed and asked for more time, since I was being deposed in two plaintiff cases and not one. This argument was not resolved until the judge agreed with Monsanto that they could depose me and they could have more time with me. We set up another deposition date. I wasn't too pleased with that, but thought" here we go again."

When I arrived for the second deposition, the doctor-lawyer was not present. But Tomaselli trotted out some more possible causes. First, he wanted to discuss Alva's ulcerative colitis. Alva had been diagnosed with ulcerative colitis when he had a colonoscopy for gastrointestinal symptoms, and he was treated with an oral medication called Asacol (mesalamine), a non-steroidal anti-inflammatory. As I've noted already, some studies had shown a potential association between ulcerative colitis and lymphoma in patients who were treated for the colitis with immunosuppressive agents, but Alva had not been on such therapy. Furthermore, subsequent colonoscopies did not show that he had ulcerative colitis, and his gastrointestinal symptoms had never followed the usual waxing-and-waning pattern seen in that disease. All of that made me question whether Alva had truly had traditional ulcerative colitis, though I acknowledged that one pathology report had shown possible colitis.

One important point had not yet been made. And as we were nearing the end of the second deposition, I wanted to highlight it. "These two plaintiffs were a married couple," I said for the record. "And this should raise a red flag as to the common denominator they were both exposed to." To me, this was common sense. When two people living in the same household develop the same disease, we shouldn't be looking only at individual factors. We need to look for something they both had in common. In the Pilliods' case, it seemed obvious: Roundup.

But that wouldn't prevent me from having to answer the same questions all over again. This time, I would have to fly to San Francisco to testify in front of Judge Winifred Smith in Alameda for what is called a "Sargon hearing." This was part of the pre-trial process, similar to the Daubert hearing I'd been through in front of Judge Chhabria. In a Sargon hearing, the judge determines whether the methodology used by expert witnesses on each side is solid, so that the experts' conclusions will be based on evidence and not merely on conjecture. I learned that the Daubert standard is applied only in federal court and in the courts of some states, but in California the courts use the Sargon standard.

My job was to explain to the court what I did and how I applied the evidence in the Pilliods' case to reach my conclusions on the causes of their lymphoma. The judge in turn was to evaluate my methodology and how I conducted my analysis so that she could rule as to whether I would be allowed to testify in front of the jury. The testimony date was scheduled for Wednesday, March 6, 2019.

I arrived in San Francisco the day before my planned testimony and took an Uber to a Marriott in downtown Oakland. There, I was reunited with Mike Miller. It was great to see him. Miller was licking his chops at the thought of taking on Monsanto in front of a jury. But this time around there would be no jury for me. There would be only a bunch of well-dressed lawyers, a judge, and whoever else wanted to attend from the press and general public.

As we drove to the courthouse the following day, Mike was his usual cheery self—confident, encouraging, and smiling warmly. "Chadi," he said in his pleasant Virginia drawl, "I know you're going to do a great job today. Now they're trying to blame it on Alberta being a schoolteacher? No jury is buying that." He winked. "Especially if we manage to have a couple of schoolteachers on that jury."

As soon as I was sworn in, Judge Smith asked Kelly Evans, yet another Monsanto attorney, to begin with the cross-examination. As I've noted, usually the direct examination comes first, but, frankly, everyone appears to know that cross-examination is way more important than direct questioning, so presumably it seemed more efficient to get that out of the way first.

Evans began by remarking that lymphoma had existed well before Roundup came on the market in 1974, and of course I concurred. He then wanted me to say how many lymphomas have an identifiable cause, but I said I was unable to provide an actual percentage. I emphasized again that while the majority of lymphomas we see have no known cause, this should not preclude us from investigating causation for every new lymphoma that we encounter in our medical practice.

I was expecting Monsanto to once again try to claim that in making my assessment I relied only on the IARC report, which was untrue. Evans did not disappoint:

Kelly Evans: *Now, in looking at your report, you rely heavily on IARC's assessment of glyphosate, correct?*
Chadi Nabhan: *It's one of the things I relied on, but I also read pretty much every epidemiologic study I could get my hands on. I actually looked at the AHS, too. which I'm sure we're going to go through.*

Evans didn't like that answer and tried to interrupt me—I think he probably wanted to try rephrasing the question in order to get a different answer out of me, one that might better suit the narrative he wanted to establish—but the judge halted him and told me to keep speaking.

CN: *I just don't know what you meant by "heavily." So, maybe you can explain that to me.*

KE: *I think you've been asked that before and you answered that yes, you relied heavily upon the IARC. Correct?*

CN: *I have relied upon IARC, amongst other things.*

Seemingly not getting the satisfaction he wanted, Evans moved on to discuss my opinion about age as a risk factor. "Here we go again," I thought. "Let's blame the old people for being old!"

KE: *I want to just make sure I understand. With respect to age, is it your opinion that age is a risk factor for the development of NHL or not?*

CN: *Age is a risk factor for the development of every cancer under the sun including NHL because cancer is a disease of the elderly. What I said is I don't believe age is a causative risk factor. So, I don't believe age by itself causes cancer.*

KE: *And just so I make sure I understand. It's a risk factor, but it's not a cause; is that what your testimony is?*

CN: *I don't believe age causes cancer. I believe it's a risk factor to developing cancer amongst all other diseases.*

Another thing Monsanto had repeatedly highlighted was "idiopathic" lymphoma—what I've called the "we don't know" defense. In fact, George Lombardi had emphasized this in his cross-examination of me in the Johnson trial, and clearly the jury hadn't found that very convincing. You would think that Monsanto would have given up on this argument by now, but nope—Evans decided to try again.

KE: Now, how did you rule out that it was an unknown cause that resulted in Mrs. Pilliod's cancer?

CN: Because there are known causes that these cancers have. Idiopathic, by definition, is when you can't find any known cause that is implicated in the disease of this particular patient. So, if you crossed every single thing out; then that is idiopathic, then that is unknown.

KE: And as soon as you identify anything that you consider to be a cause, you say any unknown cause did not play a role in this person's cancer; correct?

CN: Because you already find a known cause—

KE (interrupting): Is that correct?

CN: Yes, you already found a known cause. So, you can't say I'm going to ignore the known cause and assume it's unknown.

KE: But you're not doing just that; you're also dismissing everything else that goes on.

CN: I didn't dismiss anything.

KE: Sure, you did. You crossed out "unknown cause."

Was he kidding? I cast a glance at Mike Miller, hoping he would intervene to end these semantic games. But as I was learning, semantic gamesmanship can be one key to success in a trial.

Evans continued down this path. "There could be a person, including someone sitting in this court today, who is already way down the path towards getting NHL, and it may be diagnosed, you know, next year; right?"

What was he trying to say? After so many back-and-forth exchanges, I thought I needed to put this discussion to rest.

CN: *Can I just give you an example to simplify all of this? Take a patient who has a heart attack and who smokes, but that's the only risk factor. You go through hypertension, diabetes, everything else. Are you able to tell this patient that "Okay, I think you probably had an idiopathic heart attack, and I'm going to ignore the fact that smoking is a known risk factor for heart attack"? Or are you going to say, "I looked at everything and I think smoking may have contributed to your heart attack"?*

KE: *But in those individuals who don't have other risk factors, other causative factors, you're going to say Roundup was the cause?*

CN: *I will not dismiss Roundup because I know it's a cause of NHL. And if somebody is exposed to it and had NHL, I think dismissing a causative factor, you would not be doing differential etiology the proper way.*

KE: *So, you're going to say, again, for that person, one of the 75,000 people in 2019 who are going to get NHL anyway, if they're exposed to Roundup, you're going to assign the cause and you'd be wrong in that person, correct?*

CN: *I don't agree 100 percent with what you're saying. I said I will look at every case individually, and if there's a causative factor that this person is exposed to, I can't dismiss it. If I have a patient that was going to get non-Hodgkin lymphoma anyway and they are HIV positive, what do I say? "You know what, you're going to get it anyway, it's not HIV positivity"? So that example you provided applies to every single etiologic factor. Why would I treat other viral associations or autoimmune diseases differently? Why would I say for the ulcerative colitis—which I did not cross out, by the way, in Mr. Pilliod—why would I say it has nothing to do with it? But that's how we do it. We put everything together and then go through the process.*

But nothing seemed to make Evans move on from this topic. The judge finally stepped in, saying, "I think you're going to have to agree to disagree."

Evans turned next to Alberta Pilliod's employment. I told him flat out that I did not consider school teaching to be a cause of her lymphoma, and that the idea was, quite frankly, nonsensical.

KE: Even though you were shown at your deposition that there is peer-reviewed literature showing that schoolteachers, for example, are at increased risk for NHL?

CN: I think we both know that I disagree with the conclusion of that paper.

KE: So, again, it's one thing to say, "I'm going to consider these factors and rule them out," but in your analysis in your report, you actually did not even consider those things; correct?

CN: Again, I said I would consider the things that I believe contribute to the risk of NHL as risk factors, whether they're causative or not. I don't believe school teaching is a risk factor for NHL, and so I'm not going to include it.

As if he was following a checklist, Evans then moved on to Alva Pilliod's history of skin cancers. Why hadn't I considered that?

"For the tenth time," I replied, "I would only consider the factors I believe contribute to the development of NHL. I don't think basal cell cancer or squamous cell cancer increase the risk of NHL."

Next, Evans brought up Mr. Pilliod's other medical conditions.

KE: So even though he had these different medical issues, that does not mean he had any kind of a weak or compromised immune system?

CN: I don't believe that this constellation of diseases you described caused his NHL, no.

KE: That was not my question, sir. My question was: You put these three things together, you don't believe that shows he had a weakened immune system?

Miller objected, pointing out that I'd already been asked that question several times, but the judge overruled him and asked me to answer. I started to, but Evans clearly did not like where I was going, for he interrupted me and said firmly, "Please answer my question, sir." So, I did, explaining that Alva's medical problems did not imply true compromise in his immune system.

At that point Evans appeared to have had enough of me, and he concluded his cross-examination by thanking me. I looked over at Judge Smith, who remained expressionless.

Now it was Mike Miller's turn for direct examination. He brought up a new analysis that had just been published a couple of weeks earlier, in which the authors showed an increase of over 40 percent in the risk of lymphoma in patients who were heavily exposed to Roundup.[4] Interestingly, these authors had once served on the Scientific Advisory Panel for the EPA and did their own analysis. Mike then moved to address all the elements that Evans had brought up—beginning with teaching school.

Mike Miller: *Is there the slightest chance in your mind that you think teaching school increases somebody's risk of non-Hodgkin lymphoma?*

Chadi Nabhan: *Of course not. I don't believe anybody in this courtroom believes that being a schoolteacher increases the risks of NHL.*

Mike brought up websites maintained by Moffitt Cancer Center and the City of Hope National Medical Center, the institutions that employed two of the experts retained to defend Monsanto in this case.

MM: *Are you aware that on the defense expert's medical institutions' websites, they show pesticides cause non-Hodgkin lymphoma?*

CN: *They do talk about pesticides. They don't talk about school teaching. I don't believe any of their experts would say that schoolteachers have increased risk of NHL.*

After three and a half hours of being on the stand, I was excused, and the judge thanked me for my testimony. Evans then asked the judge to exclude me from testifying due to flawed methodology—I guess because I did not consider school teaching as a risk factor, among other things. The judge denied Evans's request; I would be allowed to testify.

"That means they'll have another chance to poke holes in your testimony," Mike told me later. "But don't worry, we'll be ready for them."

"Oh boy," I said sarcastically. "So basically, they'll ask me the same questions in a different way, and try to get me to contradict myself, or say something I don't really mean."

Mike laughed out loud. "Chadi," he said, slapping me on the back, "you're learning how the game is played."

15

The Third Trial Begins

Opening statements in *Pilliod et al. v. Monsanto* were heard on March 28, 2019. Familiar faces Brent Wisner and Michael Miller were the lead counsel representing the Pilliods. Tarek Ismail and Kelly Evans were the lead counsel representing Monsanto. Ismail would be the one questioning me, and I braced myself to go against one of the best. There was no doubt that this case was high profile. Given the media attention to the Roundup cases, I doubted you could get much higher profile.

The Pilliods' case began in state court in California shortly after a jury had handed the agricultural behemoth a loss in the federal Hardeman case. What were the company's executives thinking after that $80 million judgment against them, and the prospect of yet another one looming? As not one Monsanto executive had yet appeared in court, we had no idea what was going through their minds behind closed doors back in St. Louis. But to outsiders at least, this trial certainly looked like do-or-die for Monsanto.

Also, the fact that this would be the third trial against Monsanto in less than twelve months was impressive. These cases had been heard with remarkable speed, considering the often deliberate pace of the justice system. In less than one year, the entire world had become more familiar with Monsanto and its approaches and philosophies, as well as the company's denials and its apparent unwillingness to acknowledge any responsibility for the suffering of patients affected by their products.

Wisner hit that point—and many others—in his spectacular opening statement that day. "Monsanto keeps telling the world that this stuff is safe, that it doesn't cause cancer," he told the jury of twelve. "When you look at the evidence objectively, no one agrees with them."

Brent even gave me a shout-out in his opening statement. "You're going to hear from Dr. Chadi Nabhan," he told the jury. "He is a physician, an oncologist, from the University of Chicago. He had the privilege to meet both the Pilliods and review their case and tell us whether or not their cancer was likely caused by Roundup or something else."

Wisner focused on the fact that both plaintiffs had been unaware of the possible carcinogenicity of the product they used. Going back to a theme he had successfully used during the Lee Johnson case, he argued that Alva and Alberta were essentially denied a choice of whether to use a toxic product or not because they didn't know it *was* toxic. And he pointed out that the odds of husband and wife being diagnosed with the same type of lymphoma were vanishingly small.

In his opening statement on behalf of Monsanto, Ismail argued that both Alva and Alberta had many other medical problems. He highlighted prior cancers, autoimmune diseases, and their history of smoking. He argued that these confounding medical problems would make it impossible to accurately determine the cause of the couple's cancers. He also repeated some of the now-familiar Monsanto arguments—for example, that the EPA had determined that glyphosate was not carcinogenic.

Ismail mentioned me by name as well in his opening statement. He wanted to prime the jury against me even before they had ever met me. "They are going to call Dr. Nabhan, who looked at all the medical records of both Mr. and Mrs. Pilliod," said the Monsanto lawyer as he pointed to the witness chair. "And he is going to admit from that chair that there is nothing in the medical records that specifically rules in, that specifically identifies Roundup as having anything to do with their cancer. There is not a test, imaging study, a laboratory value, nothing whatsoever in any of her medical records or Mr. Pilliod's medical records that will identify Roundup."

Ismail also tried to suggest that neither I nor another plaintiff expert, Dr. Dennis Weisenburger, really believed what we were telling them—he implied that we said one thing to juries when on the record and another in private. "They're going to tell you they interact with other doctors at their hospitals, but when they're talking to the other doctors at their hospital, they've never told one of their colleagues that Roundup causes non-Hodgkin's lymphoma," he said. "They're going to tell you that they've both been involved with teaching medical students," he continued. "They're going to admit that they've never taught medical students that Roundup has anything to do with NHL." Ismail moved on to his punch line: "And so, as you listen to their testimony, you can consider those facts when weighing how much to put on their opinions that they're offering you in this case." Ismail was reminding the jury that the burden of proof rested on Miller, Wisner, and their team, not on Monsanto.

The first witness called to the stand on April 2 was Dr. Christopher Portier, who had testified in prior trials. Portier, as usual, was able to articulate his position in a way that captured the jury's attention and to describe the principles behind animal studies in a simple, clear way:

Brent Wisner: *Okay. What are animal studies?*

Christopher Portier: *Basically; the concept is this. If you see that a compound can cause cancer in an animal, your concern about it causing cancer in humans is much higher. If the animal is a mammal, it's much higher. So, what animal studies are, typically they're in rats and mice. They're exposed for a large portion of their life span. And at the end of that period, they're examined to see if they've gotten any cancers and they're compared against animals that are not exposed so that you can figure out whether the chemical is causing cancers in these animals.*

Wisner wanted to provide the jury with more clarity on why rats and mice are used for these studies.

BW: *Why do we use mice and rats when we're looking at issues like cancer?*

CP: *There's a lot of reasons. First of all, they are very similar to humans in the makeup of their DNA; they're very similar to humans in the biochemistry of what's happening in the body. There are clearly differences. Rats and mice are not humans. But they're similar enough that they can be used as bellwether animals to test hypotheses about dangerous toxic chemicals. They don't live as long as humans. So, they're shorter-lived, which makes them good for a laboratory experiment. If we used dogs or cats, it would be much longer experiments because they're very long-lived. These are very short-lived animals.*

After reviewing a series of animal studies, Wisner eyed the jury as he asked Portier the most important question: "And in your scientific opinion, seeing all these tumors and all these characteristics we just discussed in all these different studies, do you have any doubt about whether or not glyphosate induces tumors in animals?"

Portier's answer was unequivocal: "I don't have any doubt whatsoever that glyphosate induces tumors in animals."

The jurors were taking notes; a few in the jury box nodded.

Despite Ismail's adroit questioning, Portier did a good job withstanding the cross-examination. He emphasized the importance of looking at the entire body of evidence. He explained to the jury that he was not relying only on animal studies, but rather on these plus genotoxicity studies and the epidemiological literature. In the end, it's about the totality of evidence, and not one piece versus another.

Dr. Charles Jameson, a retired toxicologist, followed on April 4. Jameson had served on the IARC committee as a voting participant when glyphosate was found to be a probable human carcinogen. Wisner made sure the jury heard that Jameson had spent thousands of hours assessing the carcinogenicity of various compounds on behalf of IARC. He'd done all that as a volunteer; when doing IARC work, participants do not get paid except for a small stipend and coverage for their travel expenses and lodging.

Jameson explained how the IARC assessed carcinogenicity, including the methodology and the agency's classification system. The jury had heard about IARC, but it was essential that they understand the process, and Jameson delivered. At Wisner's direction, Jameson walked the jury through how IARC reached its conclusions.

Brent Wisner: *So, earlier with Dr. Portier, we put together some thing called the three pillars of causation. The first pillar was the animal data. The second one was epidemiology. And the last one was the cell data, or I think we called it mechanistic data.*

Charles Jameson: *Mechanistic data; right.*

BW: *So, did you, as part of your work on the working group, look at all three of these pillars?*

CJ: Yes.

BW: All right. I'm going to go through each one. What was the classification given to the animal data?

CJ: For the IARC monograph review, the classification for animals was that there was sufficient data that glyphosate caused cancer in experimental animals.

BW: And what does "sufficient" mean?

CJ: "Sufficient" meaning that a positive association has been seen in multiple studies, in multiple sexes, multiple species, or to an unusual degree, the tumor incidence is at an unusual degree observed in the animals.

BW: And is that the highest classification?

CJ: That's the highest classification for the animals: "sufficient."

Next Wisner moved on to the IARC's classification of the evidence from the epidemiological studies as "limited." The IARC definition of "limited" meant that "a positive association has been observed between exposure to the agent and cancer for which a causal interpretation is considered by the working group to be credible, but chance, bias, or confounding could not be ruled out with reasonable confidence."[1] Jameson explained to the jury that this meant that the association is credible once the totality of evidence has been taken into consideration along with all the studies and meta-analyses that had been performed.

The next topic they covered was the mechanistic data, or data obtained from studies carried out at the cellular level. Wisner asked Jameson about what the IARC members had theorized regarding how glyphosate causes cancer. Jameson replied, "Basically, they identified genotoxicity, which refers to affecting the DNA. Glyphosate causes extreme oxidative stress in some cells. And oxidative stress is a known mechanism for the development of cancer in humans and in animals. And I think oxidative stress has also been associated with non-Hodgkin lymphoma formation in humans."

Wisner then went for a knockout punch:

BW: *The decision to classify glyphosate as a class 2 probably human carcinogen, was that a unanimous decision?*

CJ: *Yes. It was a unanimous vote for the classification; yes.*

Wisner shifted his gaze to the jury as he asked once again, "So, every person on that working group voted yes?"

"Yes," replied Jameson.

After that, Wisner changed topics and started to ask Jameson how he'd gotten involved in this litigation.

"You want the whole story?" Jameson asked.

"Yeah, sure."

But at this point the Monsanto attorneys asked to approach the bench, and a sidebar—a confidential conversation between the judge and the attorneys, one that the jury cannot hear and which is not on the record—took place. At the end of it, the judge ultimately did not allow Wisner to pursue that topic, and so Jameson never answered the question.

Dr. Beate Ritz, one of the star witnesses in the Hardeman trial, took the stand on April 8. Ritz has always demonstrated total command of the courtroom and has always connected well with juries and judges. I wish I was half as good as she is.

In her testimony under direct examination, Ritz explained the epidemiological studies to the jury in a simplified way. She focused in particular on one that had recently come out showing a substantially increased risk of lymphoma in patients heavily exposed to Roundup—the same study I'd addressed in my testimony in the Sargon hearing a few weeks earlier. She emphasized that the study had identified an increased risk of the exact type of lymphoma that both Pilliods suffered from. The jury appeared to be paying attention to every word. And she also took the opportunity to illuminate the flaws in the AHS, as a way of explaining how the concepts of statistical significance and clinical significance can be different and why it's important to look at the totality of the evidence.

Evans cross-examined Ritz, starting out by trying to minimize Ritz's accomplishments and picking apart her resume. That did not seem to go over well with the jury, as a few were shaking their heads as Evans pursued this line of attack. Evans also used one of the oldest tricks in the Monsanto playbook by highlighting that Ritz had never looked at glyphosate carcinogenicity until she was retained to serve as a plantiff's expert witness. And Evans criticized Ritz because she had never communicated her critiques about the AHS to the study advisors themselves; Ritz replied artfully that sometimes no one could change the course of an ongoing study even despite the best of intentions.

Dennis Weisenburger testified on April 9, 2019, despite the fact that he seemed to be suffering from a bad cold and probably wasn't feeling his best. Still, Dennis's background has always been a huge asset for patients suing Monsanto. Not only is he a hematopathologist and a world expert in lymphoma pathology, but he has also studied pesticides' association with lymphomas for more than two decades.

Dennis explained to the jury the mechanisms by which researchers believe that cells become cancerous, including oxidative stress and direct damage to cells' DNA, and showed how Roundup can be implicated in such mechanisms. He then reviewed all the epidemiological studies, including the AHS. Dennis recounted the levels of exposure that both plaintiffs had had to Roundup and told the jury that neither one of them had worn protective gear, as they'd had no suspicion that Roundup could cause any problems. Dennis did mention Mrs. Pilliod's medical history, including her experience with superficial bladder cancer, but he quite eloquently told the jury how he had concluded that Roundup was the cause of her PCNSL. Then he delved into Mr. Pilliod's history, including his long-ago episode of ulcerative colitis, but cleverly explained to the jury why Roundup was also the most substantial contributing factor in Alva's lymphoma.

Weisenburger's testimony seemed to be going in favor of the plaintiffs until Tarek Ismail started his cross-examination. Ismail's style was to demand yes-or-no answers to his questions. We all know that a black-or-white answer is not always possible, especially in litigation matters, because oftentimes the witness needs to put the answers into context. However, Ismail was insistent on the witness replying just yes or no, and Judge Smith did not object.

In the discussion of Mr. Pilliod's ulcerative colitis, Weisenburger explained that it was more the treatment of the disease than the disease itself that that was a risk factor for lymphoma. He stated his opinion that the drugs used to treat ulcerative colitis could cause genetic damage and alter the immune system in a way that could lead to the development of lymphoma. Ismail then asked the court's permission to show an article to the jury that had raised the question as to whether ulcerative colitis was a risk factor for non-Hodgkin lymphoma.[2]

Tarek Ismail: You're familiar with this paper; correct, sir?
Dennis Weisenburger: Yes; I am.
TI: And it's a study of various autoimmune diseases to see whether they're associated with non-Hodgkin lymphoma; correct?
DW: Yes.
TI: And it's done in Sweden, where other witnesses have told us they have a good cancer registry; right?
DW: Yes.

Ismail then asked Weisenburger to look at one of the tables in the paper where there was a list of various autoimmune diseases and how they increase the risk of non-Hodgkin lymphoma.

TI: And indeed, in this paper sir, they don't say what you did— which is that this is a risk factor only because of the treatment for ulcerative colitis, correct?

DW: No, but it's an article that is talking about all kinds of different autoimmune diseases so they're not going to discuss each one individually. I can tell you that I looked at this literature, and if you look at the literature on ulcerative colitis prior to the use of these therapies, there's no increased risk. And if you look at the literature after the introduction of these therapies, there is an increased risk. So, most people think that this increased risk is due to the treatment and not the actual disease itself.

TI: Are you through?

DW: Yes.

TI: Great.

But Ismail himself was not done; in fact, he was just getting started. Ismail reminded Weisenburger of his own statement that a pathologist could not determine whether Roundup caused a particular lymphoma simply by examining the malignant cells under the microscope, and of his statement that genetic damage is often required for lymphoma to develop. Ismail was clearly driving at something.

TI: Now, you have actually published that some specific genetic mutations have been associated with herbicides; correct?

DW: Yes; we published some literature on the translocation (14; 18) being associated with herbicides.

TI (cheerfully): Okay; well; that's exactly the paper that I wanted to talk about.

He was referring to a 2006 paper, of which Dennis was the senior author, that looked at the risk of non-Hodgkin lymphoma with exposure to various pesticides. The paper suggested an association between lymphomas triggered by pesticides/herbicides and the presence of a particular type of chromosomal translocation. A translocation means that some genetic material moves from one chromosome to another; this might lead to tumor development.[3]

TI: And what you did in the study was you had individuals who were exposed to various pesticides, and you looked to see for this specific chromosome translocation, was it positive or negative and compared that to people who were not exposed; right?

DW: Correct.

TK: And what you did, well, gee, is this particular genetic mutation one that predicts whether someone is at an increased risk from a pesticide exposure or not; correct?

DW: Well, we didn't look at it from that perspective. We correlated the presence of that translocation with exposure to pesticides. I don't think it's predictive because lots of people get the same translocation who are not exposed to pesticides.

As they went back and forth on this topic, Ismail's questioning was superb, and Dennis was getting increasingly uncomfortable. Then Ismail asked Dennis to tell the jury whether either Alva or Alberta Pilliod showed this translocation. Dennis responded that Alberta did not show this translocation, and that doctors had not checked Alva's cells for it.

Ismail appeared to have won that exchange with Weisenburger. The plaintiffs needed to show something to the jury that would play in their favor. Brent Wisner's video deposition of Michael Koch, a Monsanto executive who had supervised company toxicologist Dr. Donna Farmer, was a logical next step. Koch performed poorly—he had few convincing answers when shown internal Monsanto emails about planned ghost-writing and a PR campaign to defend glyphosate in the face of the IARC's classification of it as carcinogenic.

One of the most important emails shown to the jury as evidence was from a Monsanto regulatory liaison, Daniel Jenkins, to other executives, including Koch, in which they discussed how the IARC might be evaluating glyphosate. In one of the responses, Monsanto's Dr. William Heydens referred to a "$1B question" (see figure 26). When Wisner questioned Koch about this, he answered that the comment was mere hyperbole. This deposition seemed to seriously damage Monsanto.

The jury was also shown a video deposition of Dr. Daniel Goldstein, medical director for Monsanto. (He was the physician who had ignored Lee Johnson's inquiry about glyphosate.) They were shown evidence that he and Dr. Farmer both engaged in

```
From: HEYDENS, WILLIAM F [AG/1000]
Sent: Thursday, January 15, 2015 4:26 PM
To: KOCH, MICHAEL S [AG/1000]
Subject: RE: EPA Glyphosate

Yes, I am sitting here pondering this as we speak....

The $1B Question is HOW could it impact?  Actually cause them to re-open their cancer review and do their
own in-depth Epid evaluation?  This is getting huge after what we heard on our call this morning.....

-----Original Message-----
From: KOCH, MICHAEL S [AG/1000]
Sent: Thursday, January 15, 2015 4:21 PM
To: HEYDENS, WILLIAM F [AG/1000]
Subject: FW: EPA Glyphosate

Re IARC, precisely what we didn't want to hear about impact, eh?

-----Original Message-----
From: JENKINS, DANIEL J [AG/1920]
Sent: Thursday, January 15, 2015 4:05 PM
To: HEYDENS, WILLIAM F [AG/1000]; LISTELLO, JENNIFER J [AG/1000]
Cc:

Subject: EPA Glyphosate

All:

Spoke to EPA:

They delayed the PC because they are still addressing some things for human health (breast milk) as well
as monarchs.

They are interested in our breast milk assay- what's the status?  When could we meet w them to discuss?

They are interested in anything we can supply re monarchs, and we are working on this now.  I imagine we
could meet with them next month to properly characterize glyphosate's impact as well as our plans to
increase populations.

IARC- they are sending delegates (trying to get names) that are knowledgeable re gly from EDSP and
oncogenicity standpoint.  The findings by IARC would likely be impactful on their analyses.

Finally, I don't think the PC would start until the end of FEB at the earliest, more likely March.

Thanks
```

Figure 26: Email from Jenkins and the responses from Koch and Heydens.

attempts to smack down any science that put glyphosate in an unfavorable light, and that they'd even joked about it in email exchanges, likening it to a carnival game, with comments like, "this is like playing Whack-a-Mole. . . . We'll be working on this, too. Isn't freedom of speech wonderful?"[4] Listening to all this, I thought that any reasonable juror hearing this in court would question the honesty of Monsanto and how they handled the scientific evidence.

Dr. Charles Benbrook, an agricultural economist, testified on behalf of the plaintiffs. Part of his testimony involved a paper he'd recently written discussing how the IARC and the EPA had come to such different determinations about the carcinogenicity of glyphosate and why the IARC's determination should be considered more trustworthy than the EPA's.[5] Monsanto tried to block his testimony at the last minute, but Judge Smith wasn't having it. Even as Benbrook was testifying under direct examination, the Monsanto attorneys were objecting frequently, claiming that the jury appeared a bit uncomfortable. The continuous objections did not seem to sit well with the jurors, though I also wondered whether all those objections had interfered with the jury's understanding of the points Benbrook was trying to make.

All this, and especially Ismail's trenchant questioning, was making me nervous about my upcoming testimony—which now was not far away.

16

Another Day in Court: The Pilliods vs. Monsanto

On April 19, 2019, a couple of days before my trip to Oakland to testify in the Pilliods' case, a peer-reviewed article about their specific form of lymphoma, DLBCL, was released online in the journal *JAMA Network Open*.[1] This monthly open-access journal is one of many prestigious publications of the American Medical Association. When something comes out in one of these publications, people pay attention. And I can tell you, a lot of people involved on either side of the Monsanto case paid attention to this study.

The article found that treatment failure rates were higher in DLBCL patients who had been exposed to pesticides in agricultural work compared to patients who hadn't had any such exposure. Significantly, the authors started their investigation with the premise that pesticides are a known risk factor for non-Hodgkin lymphoma. The paper also alluded to glyphosate and the IARC's determination that it was a probable human carcinogen. As soon as I finished reading the paper, I emailed a summary of it along with the full publication to Mike Miller and the trial team.

A few days later, as my flight soared through sunny skies from Chicago to San Francisco, I began to think: What were the odds of winning a third trial? The idea of a settlement had crossed my mind. In fact, after Monsanto lost the Hardeman trial, Judge Chhabria had postponed the second federal trial and ordered both parties to start the mediation process. Basically, the sense was that both parties need to start a dialogue that might lead to settlement, instead of continuing with more trials.

However, this decision did not affect the Pilliods' trial, for which a date had already been set in state court. So once again I was on my way to California to testify for the Pilliods. Shortly after I arrived I met up with Mike Miller, his wife, Nancy Guy Armstrong Miller, and Jeff Travers at the Marriott in downtown Oakland. The four of us discussed strategy while sitting around a large table covered with case files, scattered papers, binders, and court documents. After we shared a delicious dinner, I retreated to my hotel room around 8:30 p.m. but didn't go to bed until 1 a.m., and then I tossed and turned most of the night. You would think by the third time around I'd have been more at ease in front of a jury, but I still dreaded it.

As I took the stand on the morning of April 22, 2019, I glanced at the jurors, trying to read their expressions. After all, they were the most important people in that courtroom. If they agreed with what the other experts and I were saying, we would win the trial. If they did not, we would lose. It was that simple. The jurors in the first two trials had been a bit of a blur to me, likely because of my nerves. But this time around I noticed a few details about them. Most of them appeared relatively young; one of them, a man with shoulder-length hair, looked young enough to be a typical college student. Another one looked half-asleep. "He'd better perk up when we start talking," I thought.

As always, my time on the stand started with Mike establishing my bona fides for the benefit of the jurors. He brought up my publications, including papers on PCNSL (the disease that Mrs. Pilliod had) and lenalidomide (the chemotherapy drug that she was still taking); I also commented that I would be traveling to Europe in a few weeks to attend a hematology meeting, where I would be moderating a roundtable discussion on lymphoma. Then Mike moved to admit me as an expert in this trial. Both times before, the opposing counsel—George Lombardi in the Johnson case and Brian Stekloff in the Hardeman trial—had agreed immediately. But this time, the opposing counsel, Tarek Ismail, asked for a chance to question me. He was, essentially, challenging my qualifications. Was this a Monsanto strategy to shake my confidence? If so, it was working.

Ismail started by asking me about my new job at Aptitude Health, where I had started a couple of months earlier as chief medical officer. He cited the description of the company on the Aptitude Health website before going through how I outlined my job description and my responsibilities on my curriculum vitae.

Tarek Ismail: *Now, Aptitude Health, by and large, your clients are the drug companies, pharmaceutical companies; correct?*

Chadi Nabhan: *And the oncologists. We have essentially two major clients, oncologists and manufacturers of oncology products.*

TI: *Right. And so, for example, this bullet point here is describing how your company that you work for helps write medical articles and provide content for drug companies; correct?*

CN: *And for oncologists, again. So, oncologists do participate in the research that we do.*

TI: *The third bullet point, describing what you do: "consistently demonstrate an aptitude for analyzing market dynamics, evaluating the challenges facing a specific brand, identifying barriers and clinical success factors, and recommending appropriate* *tactics that overcome barriers and achieve successful goals." Did I read that correctly?*

CN: *Yes; you did.*

TI: *And then you share all that work with your global colleagues; correct?*

Ismail's tone was becoming a bit more accusatory, as if he was leading up to that dreaded "gotcha" moment.

CN: *Yes.*

TI: *You develop presentations for capability/pitch presentations?*

CN: *You try to explain to the outside stakeholders, the oncologists and the manufacturers, what is it that you do, what are the products or the capabilities that you have.*

TI: *Right. So, capability pitch presentations, those are like sales presentations; correct?*

CN: *I call them capabilities and pitch. You may call them sales.*

TI: *Indeed, you did. And then you persuasively articulate Aptitude's current value proposition as a strategic partner to all client interactions consistent with your company's global strategy; right?*

CN: *Yes.*

TI: *One of the things that you do is KOL; build a KOL network; right?*

CN: *Yes.*

TI: *KOL, that's an abbreviation for "key opinion leader"?*

CN: *Yes. So, for example, the meeting that I just described; I am going to moderate a meeting with KOLs in the EU.*

I was referring to the upcoming European Hematology Association meeting, at which I'd be moderating a roundtable discussion on advances in lymphoma treatment, as I'd said a few minutes earlier while Mike Miller was examining me. I got the strong impression that Ismail wanted to direct the conversation and elicit a yes-or-no answer, and also that he didn't like it when I tried to comment or elaborate. But I was going to make sure, to the extent possible, that I provided context so that my answers would be interpreted properly.

TI: *I appreciate that, but let's just define terms. So, a key opinion leader is a physician who is influential in a particular area that sometimes drug companies will turn to, to help talk about their therapies on their behalf; correct?*

CN: *Not entirely correct actually. Not just drug companies. I mean, key opinion leaders are folks who are investigators and researchers to whom oncologists turn to for their guidance. Right? So, it's not just drug companies.*

TI: *Right. Sure. But what I said is accurate. Key opinion leaders are employed or utilized by drug companies, at least in part by drug companies to do the activities I just described. True?*

CN: *They obviously are interested in their opinions.*

TI: *Yes. And so, you go out and you help recruit these key opinion leaders in part to speak on behalf of these drug companies; correct?*

CN: *Not to speak on behalf of the drug companies.*

Here Ismail tried to interrupt me, but I talked over him. I was fed up with his accusatory tone.

CN: *If you want me to explain what I do, I'm more than happy to, but you have to give me an opportunity. So, they don't actually speak on behalf of the company. They actually work with us. So, my role is to make sure I'm able to understand what is happening with EU investigators, with US investigators, because that helps me understand what happens to patients as well as to drugs being manufactured.*

The jurors seemed attentive, and Mike Miller appeared to be getting annoyed, so Ismail backed off that topic. Next he wanted to know how I looked at the animal and genotoxicity data. At that point Miller objected, saying that Ismail was no longer questioning my qualifications but rather performing a cross-examination. The judge agreed and asked Ismail to focus only on qualifications.

TI: *Okay; so, I guess we'll wait and see if you talk about animal and genotoxicity data, and I'll save that question for this afternoon.*
CN: *Please do.*

Ismail then wanted to ask me about my compensation for appearing in court, but Mike again objected, arguing that such a question had nothing to do with qualifications. Once more the judge agreed. So Ismail backed off as gracefully as he could, saying, "So, Doctor, why don't I at this point hand you back to Mr. Miller and then you and I will continue our conversation this afternoon?"

I made sure to look at the jury as I responded confidently, "Looking forward to it."

This exchange had made it clear that Ismail was trying to shake my resolve and question my credibility in front of the jury from the get-go, setting the tone for what was yet to come. But I felt now that I was ready for him. Maybe it was because by now I'd realized that as good as these attorneys for Monsanto were, they were basically working from the same playbook: asking the same questions, making the same innuendos. I was getting sick and tired of all that and wanted to assert what I felt was the truth—about glyphosate, about this company and their product, and about the good people I was there trying to help.

As if he could sense my burgeoning confidence and wanted to bolster it, Miller began his direct examination by reviewing my involvement in Roundup-related cases since 2016. I think he wanted the jury to know how I'd stood up to Monsanto lawyers before, to convey to them that I could do the same with Ismail. Referring to the times I'd been deposed by Monsanto attorneys, he said:

Mike Miller: *After twelve hours with some pretty smart lawyers representing Monsanto asking you questions, did they change your opinion that Roundup causes non-Hodgkin lymphoma?*
Chadi Nabhan: *No. Facts are the facts.*
MM: *Is it a hard call?*
CN: *Not at this point.*
MM: *Okay. So, we told you we'd pay you for your time; is that right?*
CN: *I hope everybody in this courtroom is getting paid for their time as well.*

I heard a few quiet laughs in the courtroom, and a few jurors smiled.

Miller then started discussing the new *JAMA Network Open* study I'd discovered several days earlier. I pointed out, looking at the jury as I did so, that the paper was not asking whether pesticides caused lymphoma or not; it took as a given that pesticides could cause lymphoma. Rather, the question the authors were trying to answer was whether the prognosis and outcomes of lymphoma patients who had been exposed to pesticides were worse than those of lymphoma patients not exposed to pesticides. Mike made sure the jury was aware that glyphosate was mentioned in the paper and that the authors stated in their introduction that glyphosate has been linked to lymphoma.

Furthermore, the study had included both people who were exposed to pesticides in the course of their work (what's called occupational exposure), in this case agriculture, and those who had been exposed but not as part of their work, as well as people who had not been exposed to pesticides. Significantly, the paper noted that "occupational exposure was not associated with clinical and biological characteristics at diagnosis"; that is to say, there was nothing to differentiate the lymphoma patients who had been exposed to pesticides in the course of their work from any of the other groups of patients. This was important, as Monsanto always tried to argue that the lack of clinical evidence demonstrating glyphosate exposure in lymphoma patients meant that glyphosate couldn't be responsible for their illness.

Mike Miller: *In this article that came out Friday, they said that "occupational exposure was not associated with clinical and biological characteristics at diagnosis." What does that mean?*

Chadi Nabhan: *It means there's nothing you can tell under the microscope, or no test you can do pathologically or clinically, to tell the actual cause of non-Hodgkin lymphoma. Or DLBCL, in this case. The way you do that is by getting a good history, by understanding what the patient went through. You go through risk factors for the particular disease and try to conclude whether there is a cause or there's no cause. Most often, we actually can't find a cause, and sometimes we can. So, it just tells you that at the time of diagnosis of DLBCL, there is no actual biologic marker, that you can say, "Oh, I have this biologic marker; accordingly, this DLBCL is caused by X or not caused by X." That doesn't exist.*

Tarek Ismail: *Move to strike, your honor; the last portion of that.*

MM: *He just answered the question.*

Judge Winifred Smith: *Overruled; the answer will stay.*

I then discussed the various other medical conditions that Alva Pilliod had, and I explained to the jury how I eliminated all of them as a potential cause of his lymphoma. To do so, we mounted a whiteboard to the left of where I was sitting, and I wrote on it all the risk factors for lymphoma, just as I'd done in the Johnson trial.

My plan, like in that earlier trial, was to scratch out all the factors that were not causative in this particular case, leaving only those factors that might be associated with the disease. I planned on doing that for Alberta as well.

I explained to the jury why age, sex, and race do not cause non-Hodgkin lymphoma, and given the body language I could see among the jurors, I thought my arguments were rather convincing. We then discussed obesity, ulcerative colitis, and other medical problems the couple had. I explained to the jury how none of these were substantially linked to them developing lymphoma.

Miller then showed the jury two studies that explored cases in which spouses had developed the same cancer. When he asked me for my opinion, I responded, "As a clinician, I applaud an investigation like this. But at the end of the day, it's common sense to me." When a couple who have lived together for thirty years develop the same cancer, we must investigate the possibility of exposure to the same environmental factor.

Miller concluded his direct examination a little before 3 p.m., at which point the judge determined it was too late to start cross-examination. I was unhappy with this, because now Ismail and his crew would have the entire evening to look at how today had gone and figure out how best to attack me the following morning. I asked the judge for permission to take back with me the two large binders full of scientific articles and experts' depositions that Miller had put before me, so that I could review them ahead of the following day's cross-examination, and she agreed.

By law, I was not allowed to discuss the case with the plaintiff's lawyers once they had completed direct examination. So Miller asked the judge's permission to take me out to dinner, with the promise that we would not discuss the case, and the judge approved. That evening I met Mike, Nancy, and the other members of the team at an Italian restaurant, and we stuck to our promise: we talked about the weather, about the food, about our families, and the upcoming NFL draft, with not a word about the case whatsoever.

"Good luck tomorrow," Mike said as we headed back to the hotel. "Everything will be fine."

I was reasonably sure that Tarek Ismail would do everything in his power to keep that prediction from coming true.

I woke up with back spasms the next morning and could barely get myself out of bed. I had no idea what had caused them, but it certainly was not a good omen. Three Advils and a twenty-minute shower later, I was as ready as I would ever be to face off against my adversary.

At 9:43 a.m., I was asked to take the stand. I had to lean a bit forward in my chair because of my back pain. I had on the same dark-navy suit I'd worn the day before, but a different shirt and tie. I eyed Ismail, who was sharply dressed in a buttoned dark-gray suit and looked confident and well-prepared.

At the start of his cross-examination, Ismail brought up yet another old tune Monsanto loved to sing: how the EPA had reviewed various epidemiological studies and concluded that there was no association between glyphosate and non-Hodgkin lymphoma. He highlighted the rigor of the EPA review and moved on to discuss the agency's review of animal studies. He asked if I had reviewed the animal studies myself, and I answered that I had a while back, but not recently.

Tarek Ismail: *Sure. But certainly, you did not review the animal cancer studies with the same rigor and expertise as did the EPA scientists; true?*

Chadi Nabhan: *I can't really speak of the rigor that they do. I can only say that I did not review them rigorously, but I think I can't comment on how rigorous their review [was].*

TI: *Fair enough, sir. So, they reviewed fourteen animal carcino-*
genicity studies with glyphosate, glyphosate acid, or glyphosate
salts in this 2017 review; correct?

CN: *Yes.*

TI: *And what they determined was that none of the tumors evalu-*
ated were considered to be [glyphosate]-related based on weight of
evidence evaluations. Did I read that correctly?

CN: *That was their determination.*

He then asked me if I knew that many other scientific organizations around the world (such as the European Food Safety
Authority) had examined the question of whether products
like Roundup increase the risk of non-Hodgkin lymphoma in
humans.

I tried to formulate a response that would not involve the words
"yes" or "no." I'd about had it with that tactic. How could he demand simplistic, one-word answers in complex matters like this?

CN: *Many of them have. They actually looked more at food*
contamination and whether it increases carcinogenicity in food.

TI: *And so, the answer to my question is yes?*

CN: *Yes, but it's important to clarify.*

TI: *Other organizations have examined this question about*
whether glyphosate increases the cancer risk in humans; true?

CN: *I understand but—*

TI: *Is the answer yes?*

CN *EFSA says . . . it's European Food Safety Authority; right?*

TI: *You understand—*

CN [interrupting]: *That's what EFSA—*

TI: *And you're not an expert in European regulatory system for the*
approval and registration of pesticide.

CN: *I think neither of us is.*

TI: *So, the answer is yes?*
CN: *Correct.*
TI: *Thank you, sir.*

Ismail then continued to press me about my critical view of the EFSA decision, glancing back and forth between me and the jury.

TI: *So, you told the jury that this group of scientists in Europe were wrong about this conclusion regarding genotoxicity and whether there was evidence of carcinogenicity in animal studies; correct? That's what you just said a moment ago.*
CN: *It is my opinion they were wrong.*
TI: *Right. And so, my follow-up question to you is you are neither an expert in genotoxicity nor animal cancer studies; true?*
CN: *I don't need to be an expert to know if they were wrong.*
TI: *The answer to my question is yes?*
CN: *I am not an expert. I've said that five times.*
TI: *Thank you.*

I was not sure who won that exchange. But in the chess match of cross-examination, a draw is considered a win for the person being cross-examined, and I thought I'd held my ground. I reminded myself of what Kathryn Forgie had told me during the Hardeman case: "Trials are won on cross-examination."

Ismail continued to go through a list of various scientific organizations that had looked at glyphosate and its potential to cause cancer, asking me if I knew what they had concluded. His tone was sarcastic as he continued questioning me:

TI: *And I assume you're going to tell us that it is your opinion that they got it wrong; right?*
CN: *I said I disagree; I disagree with the opinion. Reasonable people can disagree.*

TI: *And this is an issue upon which reasonable people can disagree?*

CN: *I think me and you disagree right now. So, yes, we can disagree.*

TI: *Great. And so, what you were telling us a moment ago is the question about whether products like Roundup increase the risk of non-Hodgkin lymphoma is one of those scientific questions for which reasonable people can disagree. That's what you just said; correct?*

With that, Ismail played right into my hands. I knew where I was going to go with that.

CN: *That's right, until eventually all people could agree.*

TI: *One way or another.*

CN: *Thirty years ago, people thought smoking was good.*

I looked at the jury as I said that—and Ismail interrupted immediately.

TI: *Doctor, do you remember my question?*

CN: *I remember your question.*

The jury got the important point that I was trying to make: that science evolves and substances that at one time were considered safe, like tobacco, might come to be seen as unsafe as our knowledge improves.

I knew I'd scored a point when he resorted to asking about my financial compensation.

TI: *I saw, either in your report or in your deposition, that you have an all-day rate for trials. Do you recall that?*

CN: *Yes.*

TI: *Is that $5,000 per day?*

CN: *Right. It usually takes more than 10–12 hours. So, I just set that; yeah.*

TI: You made $5,000 yesterday; correct?
CN: I didn't make anything. I haven't billed for anything.
TI: You will invoice for $5,000 for yesterday; correct?
CN: I'm not going to invoice for two separate trial days. That's not
   my plan.
TI: So, $5,000; correct?
CN: Yes; that's my plan. Unless you think I shouldn't?

I saw a few nods and smiles from the jury.

TI: You know, that's between you and Mr. Miller.
CN: I don't know; I should take notes from lawyers.
TI: You can talk about it with Mr. Miller and you can work out
   whatever you want with him.
CN: Sure.

He moved on to discuss two papers I had referenced in my
expert report and discussed during my direct examination with
Miller, ones that had investigated cases in which spouses had
developed the same cancer. Those papers had their limitations,
which I acknowledged. Again, while eyeing the jury, I said that
it was common sense to explore common factors for individuals
who live together and develop the same disease.

Ismail reminded me of when I had met the Pilliods in December
2018 in my office. He recounted how I'd asked them about their
exposure to Roundup and how their initial response was not specific enough, which made me ask them to recall how often they
sprayed and how they did the spraying. I had asked them to be as
detailed and as specific as possible, as details mattered.

TI: And you took the estimates provided by Mrs. Pilliod as part of
   that process—exposure, as it were, to Roundup—for your opinions
   in this case; correct? Do you recall that?

CN: *I have no reason not to believe the patient.*
TI: *I wasn't suggesting—*
CN: *I think it was implied.*

That was good—I could see frowns among the jurors. They didn't like Ismail's implication.

Ismail then moved on to two other hot-button issues, age and gender as risk factors. As soon as he mentioned the latter, I jumped in.

CN: *So, you're telling me that gender by itself is a risk factor for non-Hodgkin lymphoma?*
TI: *You said gender is a risk factor.*
CN: *I didn't say it was causative. I was inclusive.*
TI: *I didn't say causative. I said risk factor. Gender is a risk factor for non-Hodgkin lymphoma; yes or no?*
CN: *I can't ask you a question. I want to make sure we are level setting. What do you mean by risk factor?*
TI: *How about I phrase it the way you phrase it? Is gender a known risk factor for NHL?*
CN: *NHL is slightly more common in men than you see in women.*
TI: *And ...*
CN: *So, in my opinion it's not a risk factor. It's just more prevalent in men than in women.*

He went to the chart I had drawn during Mike Miller's questioning.

TI: *So, when you created this chart and you had a column that said "Known Risk Factors for NHL" and you put gender there, what you really meant to say was it's not a risk factor for NHL?*
CN: *It's more prevalent in men than in women.*

The jury was attentive and taking notes.

TI: *And when you said age is a known risk factor for NHL, what*
you meant to say is age is not a risk factor for NHL; right?

CN: *Counsel, I know what I meant to say. So, let me tell you what I*
meant to say—

TI (interrupting): *And then when you said—*

CN: *Okay.*

TI (interrupting): *—race and so Caucasians—*

At this point Mike Miller objected to Ismail interrupting me
(and, of course, interrupted Ismail in making that objection).

The judge said, "Okay, so everybody is interrupting everyone." Then she looked at me. "I want you to just listen carefully
to the question and respond to the question in however you want
to respond. But not everything can require either repetition or an
explanation or something that's not directly referenced to the
question. You may have an opportunity to talk about it later, but
just respond to Mr. Ismail." She then turned to Ismail and said,
"And, Mr. Ismail, don't step on the answers; thank you."

CN: *What I meant is that we see lymphoma more in older patients,*
more in white patients, and more in men. So, I put them there
because that's the prevalence of when you see non-Hodgkin
lymphoma. But it is not my belief that age or gender or race cause
the disease.

TI: *We all understand that is your view, Doctor. And with respect*
to Mr. Pilliod, each of those factors which you said put him at
an increased prevalence of getting non-Hodgkin lymphoma are
positive for him; true?

CN: *Older people are at increased risk of NHL. He was at the age of*
possibly getting NHL. We've gone through this.

TI: *And with respect to Mrs. Pilliod, we could do the same exercise. Age and race would be, in her case, factors that put her in the group that had a higher prevalence of non-Hodgkin lymphoma; true?*

CN: *And using your argument, sex would be protective then because it happens less in women . . .*

TI: *Your Honor . . .*

CN: *I'm sorry. I apologize, Your Honor.*

Judge Winifred Smith: *Thank you.*

CN: *But, I mean, the issue requires explanation. You know, it's taking things in abstract. That's why. So, my apologies.*

I was clearly not letting Ismail have the field day he might have been expecting. But I was also concerned that these exchanges were becoming acrimonious. Would that change the jury's views?

Eventually Ismail got to an argument that every Monsanto lawyer had thrown at me.

TI: *You're not aware of any imaging study that was done in Mr. Pilliod's case that would allow clinicians to identify Roundup as the cause of his particular cancer; correct?*

CN: *Correct. And similarly, there's no imaging study to say it's skin cancer that led to non-Hodgkin lymphoma.*

I was referring to the skin cancers that Alva had had. Monsanto had argued that these might have led to his lymphoma.

TI: *And there's no imaging study that was done in Mrs. Pilliod's case to identify Roundup as the cause of her non-Hodgkin lymphoma; correct?*

CN: *Imaging studies do not identify the cause.*

After more than four hours, Ismail concluded his cross-examination. I took a deep breath as he passed me back to Mike Miller for what's called a redirect examination. Miller's redirect was brief and uneventful, and just like that I was excused.

Later that day, Brent Wisner and Mike Miller announced to the judge that they had concluded their arguments. It was Monsanto's turn.

But while their lawyers would be preparing to defend them, it turned out that their stockholders were ready to pillory them.

17

The $2 Billion Judgment

While the Pilliods' trial was unfolding, the media was eyeing a much-anticipated Bayer shareholder meeting in Bonn, Germany. The CEO of Bayer, Werner Baumann, was under scrutiny for buying Monsanto for over $60 billion despite the ongoing litigation.[1]

Baumann maintained that glyphosate was safe and that Bayer would stand behind their newly acquired product. Despite losing the first two trials and being ordered to pay large amounts in damages, Bayer and Monsanto had continued to issue statements disagreeing with the verdicts.

Stockholders weren't buying it. On April 26, 2019, in what the *Wall Street Journal* termed a "strong rebuke" to Baumann, some 55 percent of the Bayer shareholders refused to endorse management's actions in the preceding year, a sharp contrast to the year before, when 97 percent had approved of management strategies. Investors had lost confidence in how the company was being run. The meeting lasted twelve hours, as anti-Monsanto protestors demonstrated outside.[2] "Management infected a healthy Bayer with the Monsanto virus," Ingo Speich, head of corporate governance at Deka Investment (which held 1 percent of Bayer at the time), told the *Journal*.

By the time the meeting was held, Bayer's share price had plummeted (in the spring of 2016, when I started working on the Monsanto cases, Bayer's share price was 103 euros, but in April 2019, when the Pilliod trial was ongoing, it was as low as 61 euros). Despite that slump in its stock, despite the criticisms from shareholders and analysts, despite the protestors, and despite the fact that they'd already lost two major trials and were in danger of losing a third, Baumann professed confidence. "The acquisition was and is the right step for Bayer," he said.[3]

I wish I could have had an hour with him to explain my views: why I thought glyphosate was harmful, what we'd learned regarding Monsanto's practices, and how badly this acquisition was tarnishing its well-respected parent company. Above all, I wished Baumann had attended one of the trials. I think it would have been eye-opening to him.

Back in the courthouse, Monsanto invited Dr. Celeste Bello, an oncologist at the Moffitt Cancer Center in Tampa, Florida, to the stand, to defend Roundup. Bello is a lymphoma specialist with a wealth of expertise in PCNSL, the disease that Mrs. Pilliod suffered from. This was her first time as an expert medical witness, so I didn't envy her when Brent Wisner decided to probe her credentials. It was, I suspect, payback for when Ismail had grilled me.

Brent Wisner: *There are people who spend their careers, experts, trying to determine if chemicals cause cancer; right?*
Celeste Bello: *There are people who do research on that; yes.*
BW: *You're not one of them?*
CB: *I would say I am as close to an expert as you'll get in that field.*
BW: *So, it's your testimony that you're as close as it comes to an expert on this area and you've never once published on it?*

CB: *Well, I don't think that's entirely fair to say.*

BW: *Okay, show me on your CV where you published the causes of lymphoma based on chemical exposure. I didn't see that in your CV.*

CB: *Well; I wouldn't say specifically on chemical exposure, but we look at causes of lymphoma in almost every article.*

BW: *Well, that's my point and that's what I'm trying to get at. There are experts who study chemicals and how chemicals cause specific cancers, and we have met some of these experts, but you're not one of those experts. You're focusing on the treatment of lymphoma; right?*

CB: *But I am an expert in lymphoma in humans and what causes lymphoma in humans, so, that's where my expertise would lie.*

BW: *Okay. But again, just to get the point, we're talking about chemicals causing lymphoma. You've never published in that area.*

CB: *No, I've never published on chemicals causing lymphomas.*

BW: *Because that's not your expertise; right?*

CB: *Not my expertise.*

BW: *At this time, your honor, we move to exclude her testimony about Roundup, as she's not an expert in the area of chemicals causing cancer.*

Judge Winifred Smith: *Overruled.*

While I certainly hoped that this might have led the jury to start doubting what she was about to say, as it would be good for our case, I couldn't help feeling sympathy for Bello. It's not easy to sit there and let someone try to minimize your achievements, rip apart your resume, and essentially belittle you.

Ismail conducted the direct examination of Bello. She said that the causes of non-Hodgkin lymphoma were largely un-known and—echoing the conclusion of every Monsanto expert—

declared that Roundup "categorically did not cause Mrs. Pilliod's disease."

Bello criticized my methodology in reaching the opposite conclusion. She also disagreed with my statement that we can use common sense when it came to determining that a common cause might be responsible when spouses got the same disease. Bello relied heavily on the AHS—whose deficits and shortcomings had been brought out in previous hearings and trials—for her assertion that Roundup did not cause lymphoma.

In conducting his cross-examination, Wisner pointed out how the AHS had concluded that the pesticide DDT did not cause lymphoma, even though we did know that it does indeed cause it. In fact, the EPA determined that DDT is a probable carcinogen.[4] The AHS was wrong on DDT, so why should we believe it wasn't wrong about glyphosate as well? Bello did not have a convincing answer. And when Wisner probed Bello on her views about DDT and lymphoma, she called it a "risk factor" but wouldn't say it caused non-Hodgkin lymphoma.

You didn't have to be a scientist to know that DDT was the poster chemical for bad pesticides—a long-banned substance that has become synonymous with the very idea of a toxic chemical. I couldn't imagine the jury not being skeptical of that claim and of this expert—which was exactly what Brett wanted.

On April 30, 2019—the day after Bello's testimony—the EPA issued a new decision on glyphosate: it did not cause cancer or any health problems as long as it was used according to its label directions, the agency declared. The media jumped on the rather odd timing of this EPA release, occurring as it did during an ongoing trial that the entire world was closely watching. "Trump EPA Insists Monsanto's Roundup Is Safe, despite Cancer Cases," was the headline in the *Guardian*.[5]

It was not clear whether Monsanto could use the new EPA guidance as evidence during the Pilliods' trial, but when I heard the news, I wondered how this would play out in court. In the EPA press release, US agriculture secretary Sonny Perdue said: "If we are going to feed 10 billion people by 2050, we are going to need all the tools at our disposal, which includes the use [of] glyphosate."[6] This decision by the EPA instantly spurred speculation among anti-Monsanto activists, with rumors flying that this decision might have been the result of Monsanto exerting pressure on the agency.

What added fuel to that fire, I'm sure—and probably caused a new round of migraines in Germany—was a revelation about the next expert witness appearing for Monsanto, Dr. Lorelei Mucci.

Mike Miller made sure that the jury knew that Mucci was an epidemiologist, not a medical doctor, not a hematologist, and not a toxicologist. He then proceeded to discuss her funding sources, highlighting an area that I'm sure made the jury sit bolt upright.

Mike Miller: *Now, you talked about your funding, the National Cancer Institute; right?*

Lorelei Mucci: *Yes.*

MM: *You're also funded by the Bayer Corporation; aren't you?*

LM: *So, one of the newest projects that we have started is a global registry of prostate cancer patients. We're recruiting 5,000 men with advanced prostate cancer, meaning they have metastatic disease already. And one of the drugs that's used for treatment of men who have metastatic prostate cancer is from Bayer. So, Bayer has been one of the funders of this particular study.*

MM: *So, the answer is yes, you're funded by Bayer Corporation?*

LM: *Well, I am actually not personally funded, but the research study that I'm working on is funded. So, I don't receive direct funding from them, but the research product is funded by Bayer in part.*

The jury's ears perked up at that; some of them were seen scribbling down notes during this exchange. Miller then decided to do to Mucci what had been done to me: get into a detailed discussion on her compensation. She admitted that she had already been paid $100,000 for her work on this trial.

Later, during his cross-examination, Mike also hit Mucci on more substantive points. She had been critical of the IARC's assessment of glyphosate, so he artfully pointed out that Mucci had relied heavily on that very same organization for sourcing in a textbook she had helped create.

MM: *Tell the ladies and gentlemen of the jury how many times you cite IARC in your book.*

LM: *I couldn't tell you the exact number. It's probably about four hundred times.*

MM: *You know, and I know that your book is on Kindle; right?*

LM: *Yes.*

MM: *And it's searchable on Kindle?*

LM: *Yes.*

MM: *And I searched it, and there were 475 references to IARC in the book.*

LM: *Right. And again, what I was trying to explain to you is that a lot of those references are because we're citing the number of bladder cancer cases, the number of colorectal cancer deaths, the number of how each specific pattern looks for these different cancers and all that data comes from IARC.*

MM: *Sure, because it's an eminently reliable and leading agency on causes of cancer in the world; that's the truth?*

LM: *It's one of the important cancer agencies that exist. It's also an incredibly important source of cancer statistics.*

MM: *And you also cited Dennis Weisenburger eight times in the book; didn't you?*

LM: *Yeah. He was a co-author on several of the early case-control studies of different cancers.*

MM: *And I don't want to be unkind, but Dr. Weisenburger has never cited you; you're aware of that; right?*

LM: *I couldn't say.*

On Monday, May 6, Dr. Alexandra Levine was called to the stand to testify on behalf of Monsanto. She had testified previously in the Hardeman trial and had clearly done well enough that Monsanto asked her to defend Roundup in this trial as well. She was being put on the stand to disagree strongly with my approach to determining causation of lymphoma and to assert that the Pilliods' lymphomas were related to their other medical conditions, age, and additional factors. Just as he'd done with Mucci, Miller poked holes in her qualifications, pointing out that Levine's colleague at City of Hope National Medical Center, plaintiff's witness Dennis Weisenburger, had a far more prolific record of research in this area than she did.

But one seemingly innocuous question that Miller asked her—about the thoroughness of her preparations and the amount of research Dr. Levine had done on this case—would come back to haunt Monsanto.

MM: *Tell the ladies and gentlemen how many hours you spent reviewing the medical records before you wrote this report.*

Alexandra Levine: *17.5 hours.*

MM: *It says "Review medical records, 7.5."*

AL: *I'm sorry. I'm sorry. "Review medical records," 7.5 hours. Sorry.*

MM: *So, you spent 7.5 hours reviewing the medical records?*

AL: *Yes.*

MM: *Thousands of pages?*

AL: *Yes.*

MM: *And then you wrote a report that Roundup was not impli-cated in any way, shape, or form; right?*

By the end of the trial, I thought that our side had prevailed in the clash of the experts. But what would the jury think? We'd soon find out, as it was time for closing arguments.

And it was also time for the latest celebrity spotting at a Monsanto trial. When director Oliver Stone showed up on May 8—the day of the closing arguments—everyone wondered if he was contemplating a feature movie or a documentary series. If so, Brent Wisner, who would be presenting the closing argument for the Pilliods, was ready for his audition.

Wisner began his closing statement by underscoring a theme he had established in both this and the Johnson case. "I said at the beginning that this case was about choice. I told you that every single person in this courtroom has the right to make a choice about what chemicals they expose themselves or their families to. And no chemical company can take that choice away from us. They can't not tell us information." The Pilliods he said, weren't given that choice. "And the reason why is because of the decisions and choices that Monsanto has made for the last forty years."

As he continued making his argument, Wisner spotlighted the testimony provided by the plaintiff's experts, including me, and compared us to the experts that had testified on behalf of Monsanto. This was where earlier questioning of Alexandra Levine came back to bite the company. "Dr. Nabhan was here to talk about lymphoma. He has treated thousands upon thousands of patients with lymphoma. And he's looked at this issue for hundreds and hundreds of hours. He testified to that," Wisner said. "To put that in context, Dr. Levine spent seven hours reviewing medical records and seventeen hours reviewing the literature. And she said, 'Oops, it doesn't cause cancer.' And that's just to give you a sense of the scope of the thoroughness of these opinions."

I believe that comparison made an impact on the jury. After all, they were putting their lives on hold for weeks to do their civic duty on this case. I think they appreciated that somebody else had put in the time to do their duty as an expert witness as well.

Wisner then moved on to what the plaintiffs were asking in terms of financial compensation. He asked the jurors to award $37 million in compensatory damages for Mrs. Pilliod and $18 million for Mr. Pilliod. Then he asked them to consider awarding at least $892 million in punitive damages, a figure that he said represented one year of profits for Monsanto—though he added that $1 billion would be a better number, alluding to Monsanto chief of regulatory science William Heydens's internal email mentioning the "one-billion-dollar question" that Monsanto needed to answer. He went on to say that he knew this was asking for a lot, but he wanted to make it clear why he was making that request. "They can afford it, and they need to pay," he said. "Because that's the kind of number that sends a message to every single boardroom, every single stockholder, every single person in Monsanto that can make a decision about the future. That is a number that changes things."

Wisner concluded on an empathetic, emotional, and genuine note: "I've carried this burden of Mr. and Mrs. Pilliod's cases for quite some time now. Now it's your turn. Let's do right by them. Let's hold them accountable." There were tears in the eyes of some of the people in the courtroom as Wisner finished. It was a bravura performance. I hoped Oliver Stone was paying attention.

In his closing argument, Ismail immediately went on the attack. "I don't think it comes as a surprise to any of you that we disagree with nearly every single thing that [Wisner] said," he began. "Conduct is not a substantial factor in causing harm if the same harm would have occurred without that conduct, which means it is the plaintiff's burden to prove that Mr. Pilliod and Mrs. Pilliod would not have developed non-Hodgkin lymphoma had they not sprayed Roundup. That is their burden."

Legally, of course, he was correct. But the way he phrased it was confusing. Ismail was telling the jury that they needed to be convinced that these patients would *not* have developed the cancers had they *not* sprayed. If I had been sitting in the jury box, I'd have been scratching my head, trying to parse that.

Ismail reminded the jury of how many expert witnesses they had listened to, how many video depositions they'd seen, and how many exhibits they'd reviewed during the five weeks of the trial. Despite all that, he said, the plaintiffs had been unable to show any evidence linking Roundup to lymphoma in these two patients. And he got in a last dig at me: "You remember Dr. Nabhan," he said with a hint of contempt, "a formerly practicing physician now turned business executive."

Ismail said that any email communications that the plaintiffs had mentioned were being taken out of context and should not be used to penalize Monsanto, as they did not represent malice or ill intent. He reminded the jury that none of the medical records showed the Pilliods' treating physicians suggesting that Roundup had caused their lymphomas. "After all this time that we've been here in this trial, the plaintiffs haven't showed you a single document or medical record or test specifically linking either plaintiff's non-Hodgkin lymphoma to Roundup," he concluded. He urged the jury not to fall for the way Wisner had manipulated their emotions, promoting "fear over science" and "emotion over evidence."

Plaintiffs always get the final say, so in his last words to the jury, Wisner was able to rebut all of Ismail's assertions. And once again he ended on a friendly, sympathetic note, suggesting he knew the jury would do the right thing: "You guys now have a big job ahead of you. Thank you so much for your time. I know Mr.

and Mrs. Pilliod are really grateful for everything you've done. And you know what? Go get 'em."

In her final instructions to the jury, Judge Smith—no doubt mindful of the anti-Monsanto protests and the beating that the company was taking in the media—made sure that the jury understood that Monsanto should not be penalized simply because it was a large corporation. "Monsanto is entitled to the same fair and impartial treatment that you would give an individual," she said. "You must decide this case with the same fairness you would use if you were deciding the case between individuals."

Judge Smith also instructed the jury how to consider the testimony presented by medical experts. "You do not have to accept an expert's opinion," she said. "As with any other witness, it is up to you to decide whether you believe the expert's testimony and choose to use it as a basis for your decision. You may believe all, part, or none of an expert's testimony." In cases where experts disagreed with each other, which often had been the case here in the Pilliods' trial, the jurors needed to weigh each opinion against the others and probe into the reasons behind each opinion. It was appropriate to compare the qualifications of the various experts, she added.

The judge concluded with the big question: money. "You may award punitive damages against Monsanto only if Mr. Pilliod and/or Mrs. Pilliod proved that Monsanto engaged in that conduct with malice, oppression, or fraud," she instructed the jurors. Then, after a few more instructions to the jury, the session concluded, with the jury's deliberations to commence the following morning at 9 a.m.

On May 13, 2019, a verdict was reached.

Judge Smith entered the room and asked the foreperson for the verdict forms, one for each plaintiff. She read both, then handed them to one of the court officers to read aloud.

Question 1: Did Roundup fail to perform as safely as an ordinary consumer would have expected when used or misused in an intended or reasonably foreseeable way?   **Answer: Yes.**

Question 2: Was the design of Roundup a substantial factor in causing harm to Alberta Pilliod?   **Answer: Yes.**

Question 3: Did Roundup have potential risks that were known or knowable in light of the scientific and medical knowledge that was generally accepted in the scientific community at the time of their manufacture, distribution, or sale?   **Answer: Yes.**

Question 4: Did the potential risks of Roundup present a substantial danger to persons when used in accordance with widespread and commonly recognized practice?   **Answer: Yes.**

Question 5: Would ordinary consumers have recognized the potential risks?   **Answer: No.**

Question 6: Did Monsanto fail to adequately warn of the potential risks?   **Answer: Yes.**

Question 7: Was the lack of sufficient warnings a substantial factor in causing harm to Alberta Pilliod?   **Answer: Yes.**

Question 8: Was Monsanto negligent in designing, manufacturing, or supplying Roundup?   **Answer: Yes.**

Question 9: Was Monsanto's negligence a substantial factor in causing harm to Alberta Pilliod?   **Answer: Yes.**

Question 10: Did Monsanto know or should it reasonably have known that Roundup was dangerous or was likely to be dangerous when used in accordance with widespread and commonly recognized practice?   **Answer: Yes.**

Question 11: Did Monsanto know or should reasonably have known that users would not realize the danger?   **Answer: Yes.**

Question 12: Did Monsanto fail to adequately warn of the danger or instruct on the safe use of Roundup?   **Answer: Yes.**

Question 13: Would a reasonable manufacturer, distributor, or seller under the same or similar circumstances have warned of the danger or instructed on the safe use of Roundup?   **Answer: Yes.**

Question 14: Was Monsanto's failure to warn a substantial factor in causing harm to Alberta Pilliod?   **Answer: Yes.**

Yes, yes, yes! This was the verdict we had all hoped for—the culmination not only of the Pilliods' trial but also, in a sense, of all of the major Monsanto cases so far. With this verdict, there was no doubt about who was culpable and why. Now we'd find out what that culpability would cost them.

Question 15: What are Alberta Pilliod's damages?
            Answer: For past economic loss, the amount of $201,166.76.
            **Future economic loss, $2,957,710. Past noneconomic loss,
                $8 million. Future noneconomic loss, $26 million.**

Question 16: Did Monsanto engage in conduct with malice, oppression, or fraud committed by one or more officers, directors, or managing agents of Monsanto acting on behalf of Monsanto?
                                Answer: Yes.

Question 17: What amount of punitive damages, if any, do you award to Alberta Pilliod?
                        Answer: One billion dollars.

23	15. What are Alberta Pilliod's damages?
24	
25	Past economic loss:* $ 201,166.76
26	
27	Future economic loss: $ 2,957,710
28	

* If liability is found, the amount stipulated by the parties for past economic damages is $201,166.76

VERDICT FORM FOR ALBERTA PILLIOD

1	Past noneconomic loss: $ 8 million
2	
3	Future noneconomic loss: $ 26 million
4	

PUNITIVE DAMAGES

16. Did Monsanto engage in conduct with malice, oppression or fraud committed by one or more officers, directors or managing agents of Monsanto acting on behalf of Monsanto?

Yes ☒   No ☐

If your answer to question 16 is yes, then answer question 17. If you answered no, stop here, answer no further questions, and have the presiding juror sign and date this form.

17. What amount of punitive damages, if any, do you award to Alberta Pilliod?

$ 1 billion

Figure 27: The verdict form for Alberta Pilliod, indicating $1 billion in punitive damages.

There were audible gasps in the courtroom. Miller and Wisner, who sat up front with Alva and Alberta, smiled triumphantly.

Alva Pilliod's verdict was next. The jury's answers were identical, and they hit Monsanto with *another* $1 billion for punitive damages in his case (see figures 27 and 28.) Two billion dollars in total—that was a sum beyond what anyone had expected. The *New York Times* called it a "staggering" award.[7]

20	**CLAIM OF DAMAGES**
21	If you answered yes to question 2, 7, 9, or 14, then answer the questions below about damages for Alva Pilliod. If you did not answer or answered no to question 2, 7, 9 and 14, stop here, answer no further
22	questions, and have the presiding juror sign and date this form.
23	
24	15. What are Alva Pilliod's damages?
25	Past economic loss:* $ 47,296.01
26	
27	Past noneconomic loss: $ 8 million
28	
	* If liability is found, the amount stipulated by the parties for past economic damages is $47,296.01
	VERDICT FORM FOR ALVA PILLIOD

1	Future noneconomic loss: $ 10 million
2	
4	**PUNITIVE DAMAGES**
5	16. Did Monsanto engage in conduct with malice, oppression or fraud committed by one or more
6	officers, directors or managing agents of Monsanto acting on behalf of Monsanto?
7	Yes No
8	☒ ☐
9	
1	If your answer to question 16 is yes, then answer question 17. If you answered no, stop here,
1	answer no further questions, and have the presiding juror sign and date this form.
2	
3	17. What amount of punitive damages, if any, do you award to Alva Pilliod?
4	$ 1 billion
5	
6	

Figure 28: The verdict form for Alva Pilliod.

Afterward, Wisner and Miller spoke with reporters. "The jury saw for themselves internal company documents demonstrating that, from day one, Monsanto has never had any interest in finding out whether Roundup is safe," Wisner said. "Instead of investing in sound science, they invested millions in attacking science that threatened their business agenda."

Miller added, "Unlike the first two Monsanto trials, where the judges severely limited the amount of plaintiff's evidence, we were finally allowed to show a jury the mountain of evidence showing Monsanto's manipulation of science, the media, and regulatory agencies to forward their own agenda despite Roundup's severe harm to the animal kingdom and humankind."

Shortly thereafter Bayer issued a statement in which they said they were "disappointed with the jury's decision" and promised to appeal. "We have great sympathy for Mr. and Mrs. Pilliod," the statement read. "But the evidence in this case was clear that both have long histories of illnesses known to be substantial risk factors for non-Hodgkin's lymphoma (NHL), most NHL has no known cause, and there is not reliable scientific evidence to conclude that glyphosate-based herbicides were the . . . cause of their illnesses as the jury was required to find in this case."[8]

In other words, Bayer was saying, blame the jury for not understanding what they were supposed to do; blame the judge for not giving them clear instructions; blame the Pilliods for whatever they had or hadn't done. Blame everything except glyphosate; blame everybody except Monsanto.

In that sense, nothing had changed.

I heard about the verdict back home in Illinois and felt, simultaneously, both an enormous burden lifted and an enormous sense of pride that science and truth had prevailed. Lee Johnson, Edwin Hardeman, and Alberta and Alva Pilliod—people apparently of little consequence to some—had gotten the justice they deserved for their suffering. By going to trial, these patients had paved the way for many others who were affected by Monsanto's product. And the dedication of the legal team and of all the expert witnesses had made this happen. Being a very small part of this entire saga was an honor that I will cherish forever.

18

Endings

I started writing this book as the world first started hearing of COVID-19. And while the pandemic may have stopped many things in their tracks, the Roundup cases dragged on—albeit remotely.

As I mentioned earlier, in September 2018 Monsanto had filed a motion for a retrial and a JNOV motion in the Johnson case (essentially calling on the judge to nullify the jury's verdict), claiming that I had not excluded an idiopathic cause. Judge Bolanos denied those motions, but she did reduce the damages awarded from $289 million to $78.5 million.

On April 23, 2019, Monsanto filed its appeal in the Johnson case, again asserting (among other claims pertaining to other experts in the case) that I ignored the possibility of idiopathy. Johnson's legal team filed their responses on May 24, 2019, debunking every one of Monsanto's assertions and urging the court to deny the appeal and even reverse the reduction in punitive damages.

While these legal fights were taking place, Mr. Johnson was undergoing therapy and decided to share his experience with the world by participating in Antlee's rap song "Not Your Time." You can watch it on YouTube. [1]

On June 2, 2020, the California Court of Appeal's First Appellate District heard oral arguments in this appeal. Because of the pandemic, all arguments were done by phone, with Monsanto represented by David Axelrad and Mike Miller representing Mr. Johnson. The appellate court usually has ninety days to make a decision, but it didn't take them that long: on July 20, 2020, they ruled in favor of Lee Johnson. The California Supreme Court refused to hear the case, and Monsanto decided not to appeal to the US Supreme Court. Monsanto ended up paying a total of $25 million to Johnson and his team, a figure that includes interest. I'm happy to say that Lee Johnson remains alive as of the date of this writing.

After the Hardeman verdict was announced in late March 2019, Judge Chhabria postponed the second federal trial and ordered mediation between the parties in all outstanding cases. He then appointed attorney Kenneth Feinberg as a mediator to attempt settling the MDL cases. Feinberg had previously been appointed as special master of the US government's September 11th Victim Compensation Fund. He'd also been the government-appointed administrator of the BP Deepwater Horizon Disaster Victim Compensation Fund. Moreover, he'd been appointed by the commonwealth of Massachusetts to administer the victim assistance fund established after the 2013 Boston Marathon bombing.

On July 3, 2019, Judge Chhabria heard arguments on Monsanto's request for a reduction in the compensatory damages and elimination of the punitive damages in the Hardeman case. On July 12, Chhabria denied Monsanto's motion for a new trial; on July 15, he issued an order stating that the compensatory damages were appropriate but ruling the punitive damages were constitutionally impermissible and reducing them to $20 million. That brought the total Hardeman award down to a little over $25 million from the original $80.27 million.

Several months later, on December 13, 2019, Monsanto filed an appeal of the Hardeman verdict in the US Court of Appeals for the Ninth Circuit. The case was argued before the appellate court on October 23, 2020; subsequently the court ruled in favor of Hardeman, denying Monsanto's appeal. Bayer, the parent company of Monsanto, subsequently filed a petition asking the Supreme Court of the United States (SCOTUS) to review the case and render an opinion. The company argued that a "warning" cannot be added to a product without EPA approval, and since the EPA has repeatedly concluded that Roundup was safe, a warning is not appropriate. Monsanto was also asking the Supreme Court to exclude expert testimonies in the Hardeman case.

Aimee Wagstaff and the rest of Hardeman's legal team were not surprised by Bayer's actions. "While paying out billions of dollars to settle claims, Monsanto continues to refuse to pay Mr. Hardeman's verdict. That doesn't seem fair to Mr. Hardeman. Even so, this is Monsanto's last-chance Hail Mary." She added, "We are eager and ready to beat Monsanto at the Supreme Court and put this baseless preemption defense behind us once and for all."

In December 2021, the Supreme Court asked President Biden's administration to provide its views. The justices generally give deference to the Solicitor General's conclusions. But, on Tuesday May 10, 2022, President Joe Biden's administration asked SCOTUS *not* to hear the Bayer case. Solicitor General Elizabeth Prelogar, who represents the administration before the high court, said in a court filing that Bayer's appeal should be rejected. On May 23, 2022, various agricultural companies wrote a letter to President Biden urging him to reverse the Solicitor General recommendations. On June 21, 2022, the SCOTUS rejected Bayer's bid refusing to hear the Hardeman case. It took over 3 years from the verdict rendered in Hardeman's favor in Judge Chhabria's court room until the Hardeman issue appears to have been resolved and has concluded. Hardeman remains alive and well, as of the time of this writing.[2]

Within a couple of months of the $2 billion verdict, Judge Smith reduced the damages and total awards to the Pilliods to a total of $87 million in an order filed on July 25, 2019, while denying Monsanto their request for a new trial. Alva's total damages were reduced to about $31 million, while Alberta's were lowered to around $56 million.

On February 7, 2020, Monsanto filed an opening brief in its appeal of the Pilliods' verdict in California's First District Court of Appeal. Because of COVID-19, the hearing was delayed until June 2021. When the hearing eventually took place, Presiding Justice J. Anthony Kline told Monsanto's lawyer, David Axelrad, that the facts presented in the company's brief had not been "evenhanded," as they failed to fairly describe the evidence not in Monsanto's favor. The judge berated the Monsanto attorney: "Mr. Axelrad, I'm rather astonished at the briefing in this case because of the distortion of the facts." Justice Kline particularly expressed his dissatisfaction at how Monsanto characterized our collective testimonies as experts defending the Pilliods. "I have never said to a lawyer who has appeared before me in the forty years I have been on this court, what I said to you," he affirmed, and added, "And I have thought about it quite a bit."[3]

The justices denied Monsanto's appeal. In their August 9 order, the judges vindicated me by criticizing how Monsanto mischaracterized my testimonies and those of other experts. "Monsanto's conclusory contentions that Dr. Weisenburger and Dr. Nabhan 'dismissed' or 'discounted' alternative causes or did not explain why they had ruled out those alternatives, are unpersuasive in light of Monsanto's failure to fairly present the substance of their testimony," the court asserted.

Monsanto's assertions that I never ruled out idiopathic causes for the Pilliods' illnesses were also rejected. "A fair reading of Dr. Weisenburger's and Dr. Nabhan's testimony does not support Monsanto's conclusion that they 'made no attempt to explain why idiopathic causes could be excluded from consideration,'" the judges wrote. They also dismissed Monsanto's assertions that I simply used "common sense" to explain causation. That was not my testimony, and the judges agreed: "Dr. Nabhan also stated that it was common sense that when two people who live together for decades develop a disease, any physician would ask whether there was a common factor between the two."

On June 27, 2022, SCOTUS upheld the $87 million judgment that was awarded to the Pilliods. The judges turned away Bayer's appeal and left in place a lower court decision upholding the judgment. Bayer's appeal in the Pilliod case argued that the verdict violates the US constitution's due process protections to award punitive damages that far outweigh compensatory damages. SCOTUS rejected that assertion. Alva and Alberta Pilliods remain alive as of the date of this writing.

In the March 2022 annual report, Bayer said that it had resolved about 107,000 cases of about 138,000 cases overall. The total number being stated in various media outlets approaches $11 billion. How these funds are allocated and how much each patient receives are details that I am not privileged to know. I suspect that there are certain parameters that both sides agreed on that determine how much each patient receives, but I wouldn't know what these parameters might be. However, it turns out that some additional Roundup cases were not part of the proposed settlement that Bayer offered; some patients might decline the settlement offers, and on occasion lawyers might advise their patients not to settle, especially if they feel that their cases can be very strong if tried in court. So you will continue to see these TV commercials from various law firms urging patients with lymphoma to call and check if they qualify.

As part of the negotiations, Monsanto tried to put an end to future potential trials, attempting to limit their future legal liability from claims associated with glyphosate that had yet to be filed (that is, future diagnoses that patients might claim were related to Roundup). Judge Chhabria pointed out that Monsanto's proposal had "glaring flaws" and that parts of the plan were "clearly unreasonable" and unfair to cancer sufferers.

Monsanto had proposed forming a "science panel" that determines whether the herbicide causes cancer or not, rather than leaving future decisions to courts and juries. Chhabria declined this request, which he said was driven simply by the fact that Monsanto had lost three trials in front of three different juries. Chhabria further wrote that Monsanto had "lost the battle of the experts" in those three trials—and just like that, Chhabria became one of my favorite people.

While all of these appeals and SCOTUS drama were taking place, Monsanto did indeed win several verdicts - four as of the time of this writing to be exact. These 4 trials were not litigated by the same law firms I had worked with. I did not serve as an expert witness in any of these four trials, nor did Drs. Weisenburger and Ritz. A fifth trial started in August 2022 in St. Louis. Whether additional trials will go in front of the jury or not is not known.

The Miller Firm is continuing to work on some cases that were not part of the proposed Bayer settlement and will be tried separately. The firm has told me they might need my opinion in some of these cases, so I might face Monsanto again at some point. Sadly, though, Mike Miller won't be a part of those efforts; he died in December 2021.

As for some of the other attorneys I worked with: Jeff Travers stays up to date with everything published on Roundup. I am amazed by his memory, and he continues to work hand in hand with the Miller Firm. David Dickens, the cool, calm, and collected attorney who put me on the stand for the first time ever in the Johnson trial, is managing the Bayer settlement and the firm's opt-out clients; he continues to fight Monsanto tooth and nail in his own calm way.

One of the saddest situations I have encountered during this litigation has been the fate of Tim Litzenburg, the attorney who dressed like he was going to a rock concert and who was with me for the first few Monsanto depositions. I grew to like Tim, but I knew something was going on when Mike Miller called me one day to tell me he was no longer with his firm. It turned out that Mike had evidence that Tim had been diverting patients with lymphoma away from the Miller Firm, but Tim had denied that claim; the two lawyers ended up suing each other. Shortly after Tim had joined another law firm, he was accused by federal prosecutors of trying to extort $200 million from an unnamed company. Tim pleaded guilty, and in September 2020, a federal judge sentenced Tim to two years in prison; he is unlikely to be able to practice law ever again.

I stay in touch with Kathryn Forgie, especially during the Premier League season. She is destined to get a text from me every time Chelsea loses and Manchester United wins. Kathryn is now pursuing claims against another pesticide, paraquat, which is claimed and alleged to cause Parkinson's disease.

Aimee Wagstaff went on to start her own law firm and continues to work with her law partner Kathryn Forgie and the Miller Law Firm on behalf of patients injured by various products and medical devices. She is working now on the paraquat cases with them. In 2013 Aimee started a group called Women in Mass Torts (WEM) with the concept of supporting and exchanging ideas with other women lawyers who are working in mass torts; WEM now has more than four hundred members.

Aimee, Kathryn, and another lawyer in their firm, David Wool, collectively won the Colorado 2019 Case of the Year Award for the work they performed on Hardeman's behalf. In 2019, Kathryn was also named to the National Trial Lawyer Top 100. I also heard that Aimee and Kathryn might still try Roundup cases. It feels that this saga is a never ending one.

Brent Wisner made many TV appearances after the Johnson trial. He has also won several awards, including the 2020 Law360 MVP of the year in product liability. He has been recognized by many prestigious organizations such as Best Lawyers, National Law Journal, LawDragon 500, and National Trial Lawyers, among others. He was featured on the cover of *Super Lawyers* magazine in 2021. Brent continues to concentrate his practice on pharmaceutical class action mass tort litigations, toxic-tort injuries, and consumer fraud litigation.

Robin Greenwald continues to work on behalf of plaintiffs injured by Roundup, as some of her plaintiffs opted out of the settlement.

* * *

As I am reminded many mornings when I walk into the gym and see yet another commercial for another law firm on the TV, the saga of Roundup continues. But finally the company did something besides filing losing appeals: Bayer announced that it will stop all US sales of Roundup for residential use and remove current versions of Roundup from store shelves by 2023.[4] It will replace Roundup with products that do not contain glyphosate, but only for American residential users; farmers will continue to have access to the glyphosate-based product. When making this decision, the company announced, "This move is being made exclusively to manage litigation risk and not because of any safety concerns," and added, "As the vast majority of claims in the litigation come from Lawn & Garden market users, this action largely eliminates the primary source of future claims beyond an assumed latency period." I believe that removing glyphosate from Bayer's lawn and garden products will protect many consumers from getting future lymphomas.

On June 17, 2022, a US appeals court ordered the EPA to take a fresh look at whether glyphosate poses unreasonable risks to humans and the environment. I thought this is a big win, as Monsanto has always argued that because the EPA determined glyphosate was safe, nothing else is needed. The San Francisco-based 9th US Circuit Court of Appeals agreed with several environmental groups that the EPA needs to look more critically at the evidence. The court felt that the EPA needs to review the evidence carefully, implying that this was not the case before. Reports have suggested that Bayer has set aside $4.5-5 billion for future settlements, but who knows if this number will go up or down.[5]

As far as me, I continue to hope that my day has more than twenty-four hours. With twin boys in high school, I suspect that navigating teenage problems could be more stressful than facing Monsanto. (Okay, maybe not.)

Being involved in this litigation brought a lot of stress, burnout, and mental exhaustion to my life. But I wouldn't change a thing. In the end, it's about the patients who suffer. They are front and center in everything we do.

The pandemic took its toll on me like it did on everyone else. So I decided to tackle some fun topics on my weekly podcast, *Healthcare Unfiltered*, including a "Docs Who Rock" episode. And I decided to learn a foreign language to challenge myself, so I take weekly lessons via zoom. For the foreseeable future however, my "Healthcare Unfiltered" podcast will remain in English, and I hope you listen to it.

ACKNOWLEDGMENTS

This book was a labor of love, but it could not have happened without the support of my wonderful family. My deepest gratitude, thanks, and love to my entire family, who tolerated my long hours, weekend after weekend of writing at coffee shops, missing family events, and the time commitment involved in countless depositions, testimonies, and preparations. Thank you to my wife, Lama; my twins, Yazan and Zane; my wonderful parents; my brother Fadi and his family; my amazing sister Dina; my in-laws; relatives; my supportive friends; and all my loved ones in the United States, Syria, Dubai, Lebanon, Germany, and my gym who were witnesses to my involvement in this litigation saga.

I want to thank the patients who brought on these lawsuits against Monsanto. You are the ones who suffered through illnesses and treatments; you are the ones who were scrutinized by Monsanto; and you're the ones who had the perseverance and determination to keep fighting until the end. You inspired me every time I read a medical record, every time I prepared for a deposition, every time I read a paper, and every time I was on the stand. You're the real heroes, and you're the reason why this book exists.

My deepest and sincerest thanks to John Hanc, the amazing writer and editor who helped me throughout these chapters and provided guidance, ideas, edits, and suggestions to make the book easier to read and comprehensible to all. His literary skills were fundamental in linking together the complex and confusing parallel worlds of science, law, and corporate culture. I am forever indebted to John for all his help and thoughtfulness. Thank you, John, for your patience, your coaching, and your help bringing this story to life within these pages.

I want to thank Joe Rusko for his advocacy for this book and for giving me the opportunity to write for the general public. Joe believed in me, in the story, and in the patients. He provided unwavering support throughout the process, and for that I am forever grateful.

I want to thank the staff at Johns Hopkins University Press, who have made this project come to life.

Special thanks to Kelly Ryerson, whom I met during these trials. Kelly has dedicated her life to environmental protection; she knows more about glyphosate and Roundup than any of us could ever imagine. She attended every trial and covered the events on her website, www.glyphosategirl.com; Kelly allowed me to reread her vivid blogs, which helped me fill in some of the gaps in this riveting story. Thank you, Kelly, for your support and for all that you do. I also want to thank Christina Liu, who is now a medical student but who edits for *The Health Care Blog*; Christina read a few of my chapters and provided much-needed comments and ideas.

My humblest thanks to all the lawyers and legal firms that I have worked with during these trials. These lawyers fought hard, put it all on the line, and defended patients against a large agricultural behemoth, even though the odds seemed to be against them.

I specifically want to thank Mike Miller, even though he is no longer with us. His firm first retained me as an expert in the spring of 2016. Mike was an amazing lawyer and litigator. He had a presence in the courtroom that few lawyers could ever parallel. It was impossible not to love Mike (unless you are Monsanto), because he was compassionate, down to earth, and passionate, and he genuinely cared about every patient he represented. The world lost an amazing human being with his death. I miss him, as many people do. I am sad and angry that we won't have twenty more years of Mike Miller around us. I want to thank him for believing in me and in my ability to represent patients that needed experts to testify and examine the evidence.

I want to thank Mike's wife, Nancy Miller, for also believing in me, and for being the amazing person that she is. Thank you to all the other members of the Miller Firm, including David Dickens, Jeff Travers, Curtis Hoke, Brian Brake, and all the Miller Firm staff.

Special thanks to the Andrus-Wagstaff law firm, especially to Kathryn Forgie, who provided me with unparalleled counsel and help during the preparations and trial appearances. Kathryn took so much time out of her schedule to explain the nuances of the legal proceedings to me. Kathryn: I still don't understand 99% of the legal stuff you taught me. She was the first to tell me that trials are won on cross-examination, a pearl that will stay with me forever. Thank you, Kathryn, for all that you do.

Special thanks to Robin Greenwald, an amazing lawyer at Weitz and Luxenburg, a legal firm that represented thousands of plaintiffs. I worked with Robin on behalf of a few patients, and I saw how deeply she cared about these patients and how committed she was. Thank you, Robin, for your help, compassion, and support.

Also, special thanks to Brent Wisner from the Baum Hedlund law firm. Brent was instrumental in winning the Johnson and Pilliod trials. His memory is unmatched, and his presence in the courtroom is out of this world. I'll say it again: Brent was born to litigate. If you have two hours to spare, watch his entire closing statement in the Johnson trial on YouTube. Thank you, Brent, for fighting the good fight and for all the encouragement you provided when I doubted myself.

Thank you to all the other expert witnesses who testified on behalf of the affected patients. Thank you for your dedication, reviews, support, and for helping patients as they fought for justice.

Lastly, thank you to everyone who believed in me and believed I could indeed write this book to chronicle the events of the past several years. Thank you for pushing me, encouraging me, and not doubting me when I doubted myself. Thank you for pushing me to get up and keep at it when I was down. I am indebted to all of you and cannot thank you enough.

NOTES

Preface

1 "Historic Ruling Against Monsanto Finds Company Acted with 'Malice' Against Groundskeeper with Cancer," interview by Amy Goodman, *Democracy Now!*, August 14, 2018, video and transcript, https://www.democracynow.org/2018/8/14/historic_ruling_against_monsanto_finds_company.

2 Michael Hiltzik, "Did a Jury Ignore Science When It Hit Monsanto with a $2-Billion Verdict?," *Los Angeles Times*, May 17, 2019, https://www.latimes.com/business/hiltzik/la-fi-hiltzik-monsanto-glyphosate-verdict-20190517-story.html.

3 "Judge Chhabria Again Turns Down Effort to Settle Bayer Monsanto Roundup Future Claims," *Corporate Crime Reporter*, May 26, 2021, https://www.corporatecrimereporter.com/news/200/judge-chhabria-again-turns-down-effort-to-settle-bayer-monsanto-roundup-future-claims/.

Chapter 1: The Phone Call

1 "Monsanto's Glyphosate Now Most Heavily Used Weed-Killer in History: Nearly 75 Percent of All Glyphosate Sprayed on Crops in the Last 10 Years," Environmental Working Group press release, February 2, 2016, https://www.sciencedaily.com/releases/2016/02/160202090536.htm.

2 Jen Monnier, "What Is Glyphosate?," LiveScience, updated September 21, 2020, https://www.livescience.com/glyphosate-round-up.html.

3 United States Environmental Protection Agency, Office of Pesticides and Toxic Substances, Archives, "Consensus Review of Glyphosate, Caswell No. 661A," March 4, 1985, https://archive.epa.gov/pesticides/chemicalsearch/chemical/foia/web/pdf/103601/103601-171.pdf.

4 United States Environmental Protection Agency, Office of Pesticides and Toxic Substances, Archives, "Glyphosate Registration Standard Revision," March 1, 1986, https://www3.epa.gov/pesticides/chem_search/cleared_reviews/csr_PC-103601_1-Mar-86_210.pdf.

5   Wayne Temple, *Review of the Evidence Relating to Glyphosate and Carcinogenicity*, prepared for the Environmental Protection Authority, August 2016, https://www.epa.govt.nz/assets/Uploads/Documents/Everyday-Environment/Publications/EPA-glyphosate-review.pdf.

6   James M. Parry, "Evaluation of the Potential Genotoxicity of Glyphosate, Glyphosate Mixtures and Component Surfactants," University of Wales, n.d.; James M. Parry, "Key Issues Concerning the Potential Genotoxicity of Glyphosate, Glyphosate Formulations and Surfactants; Recommendations for Future Work," University of Wales, n.d.; Larry Kier, "Comments on Parry Evaluation of Glyphosate and Glyphosate Formulation Potential Genotoxicity," September 18, 1999; email from Stephen J. Wratten to Mark A. Martens and Donna R. Farmer, "Subject: Comments on Parry Write-up," n.d.; letter from James M. Parry to Mark A. Martens, August 18, 1999, all presented as Exhibit 5, Case 3:16-md-02741-VC, filed March 15, 2017, https://www.baumhedlundlaw.com/documents/pdf/monsanto-documents/dr-james-parry-glyphosate-review-evaluation-of-the-potential-genotoxicity-of-glyphosate-glyphosate-mixtures-and-component-surfactants-1999.pdf.

Chapter 2: The First Meeting

1   Kathryn Z. Guyton et al., "Carcinogenicity of Tetrachlorvinphos, Parathion, Malathion, Diazinon, and Glyphosate," *The Lancet Oncology* 16, no. 5 (May 2015): 490–91, https://doi.org/10.1016/S1470-2045(15)70134-8.

Chapter 3: The EPA

1   Meir Rinde, "Richard Nixon and the Rise of American Environmentalism," *Distillations* (blog), Science History Institute, June 2, 2017, https://sciencehistory.org/distillations/richard-nixon-and-the-rise-of-american-environmentalism.

2   Lily Rothman, "Here's Why the Environmental Protection Agency Was Created," *Time*, March 22, 2017, https://time.com/4696104/environmental-protection-agency-1970-history/.

3   See Reorganization Plan No. 3 of 1970, 35 F.R. 15623, 84 Stat. 2086, as amended Pub. L. 98-80, §2(a)(2), (b)(2), (c)(2)(C), Aug. 23, 1983, 97 Stat. 485, 486 (December 2, 1970). https://www.govinfo.gov/content/pkg/USCODE-2010-title5/html/USCODE-2010-title5-app-reorganiz-other-dup92.htm.

4 "United States Environmental Protection Agency," Wikipedia, last edited April 6, 2022, https://en.wikipedia.org/wiki/United_States _Environmental_Protection_Agency.

5 Charles M. Benbrook, "Trends in Glyphosate Herbicide Use in the United States and Globally," *Environmental Sciences Europe* 28 (2016): article 3, https://enveurope.springeropen.com/articles/10.1186 /s12302-016-0070-0; Valerie Brown and Elizabeth Grossman, "How Monsanto Captured the EPA (and Twisted Science) to Keep Glyphosate on the Market," *In These Times*, November 2017, https:// inthesetimes.com/features/monsanto_epa_glyphosate_roundup _investigation.html; Securities and Exchange Commission, "Form 10-K, Monsanto Company," https://www.sec.gov/Archives/edgar /data/1110783/000111078315000230/mon-20150831x10k.htm (for net sales see bottom of page 22).

6 Brown and Grossman, "How Monsanto Captured the EPA."

7 Keith Schneider, "Faking It: The Case against Industrial Bio-Test Laboratories," *The Amicus Journal* [Natural Resources Defense Council], Spring 1983, also available at http://planetwaves.net/contents /faking_it.html.

8 See United States Environmental Protection Agency, Archives, "Roundup (Glyphosate)—EPA Reg. No. 524-308," August 9, 1978, https://www3.epa.gov/pesticides/chem_search/cleared_reviews /csr_PC-103601_9-Aug-78_060.pdf.

9 Mary Thornton, "EPA Review Finds Flawed Tests Made by Research Firm," *Washington Post*, May 13, 1983, https://www.washingtonpost .com/archive/politics/1983/05/13/epa-review-finds-flawed-tests -made-by-reasearch-firm/584839e8-8d68-4f2d-9797-decc25ecd18d.

10 Schneider, "Faking It."

11 *Federal Register* 52, no. 179 (September 16, 1987), 34911.

12 United States Environmental Protection Agency, "Memorandum: Glyphosate," June 19, 1989, https://www3.epa.gov/pesticides/chem _search/cleared_reviews/csr_PC-103601_19-Jun-89_249.pdf.

13 United States Environmental Protection Agency, "R.E.D. Facts: Glyphosate," EPA-738-F-93-011, September 1993, https://www3 .epa.gov/pesticides/chem_search/reg_actions/reregistration /fs_PC-417300_1-Sep-93.pdf.

14 Carey Gillam, "Collusion or Coincidence? Records Show EPA Efforts to Slow Herbicide Review Came in Coordination with Monsanto," HuffPost, updated August 18, 2017, https://www.huffpost.com/entry /collusion-or-coincidence-records-show-epa-efforts_b_5994dad4e4b05 6a2b0ef02f1.

15 Joel Rosenblatt, Lydia Mulvany, and Peter Waldman, "EPA Official Accused of Helping Monsanto 'Kill' Cancer Study," Bloomberg.com, March 14, 2017, https://www.bloomberg.com/news/articles/2017-03-14 /monsanto-accused-of-ghost-writing-papers-on-roundup-cancer-risk.

16 Vince Chhabria, United States District Court of Northern California, District of California, Pretrial Order No. 15, https://www .baumhedlundlaw.com/assets/Monsanto%20Roundup%20pages /Secret%20Documents/Judge-Vince-Chhabrias-ruling-to-unseal -documents.pdf.

17 "*Registration Study* definition," Law Insider, accessed April 6, 2022, https://www.lawinsider.com/dictionary/registration -study#:~:text=Registration%20Study%20means%2C%20with%20 respect,FDA%2C%20as%20described%20under%2021.

Chapter 4: Meeting Mr. Johnson

1 Steven Swerdlow et al., "The 2016 Revision of the World Health Organization Classification of Lymphoid Neoplasms," *Blood* 127, no. 20 (May 19, 2016), https://pubmed.ncbi.nlm.nih.gov/26980727/.

2 Carey Gillam, "I Won a Historic Lawsuit, but May Not Live to Get the Money," *Time*, November 21, 2018, https://time.com/5460793 /dewayne-lee-johnson-monsanto-lawsuit/.

3 Gillam, "I Won a Historic Lawsuit."

4 Gabriella Andreotti et al., "Glyphosate Use and Cancer Incidence in the Agricultural Health Study," *Journal of the National Cancer Institute* 110, no. 5 (May 2018), https://www.ncbi.nlm.nih.gov/pmc/articles /PMC6279255/.

5 A.J. De Roos, et al., "Cancer Incidence among Glyphosate-Exposed Pesticide Applicators in the Agricultural Health Study," *Environmental Health Perspective* 113, no. 1 (2005):49–54, https://www.ncbi.nlm.nih .gov/pmc/articles/PMC1253709/.

Chapter 5: The Night before the Daubert Hearing

1 "Multidistrict Litigation," Legal Information Institute, Cornell Law School, last updated June 2020, https://www.law.cornell.edu/wex /multidistrict_litigation.

2 L. Hardell and M. Eriksson, "A Case-Control Study of Non-Hodgkin Lymphoma and Exposure to Pesticides," *Cancer* 85, no. 6 (March 5, 1999): 1353–60, https://doi.org/10.1002/(sici)1097-0142(19990315)85:6 <1353::aid-cncr19>3.0.co;2-1.

3   L. Hardell, M. Eriksson, and M. Nordstrom, "Exposure to Pesticides as Risk Factor for Non-Hodgkin's Lymphoma and Hairy Cell Leukemia: Pooled Analysis of Two Swedish Case-Control Studies," *Leukemia & Lymphoma* 43, no. 5 (May 2002): 1043–49, https://doi .org/10.1080/10428190290021560.

4   H. H. McDuffie et al., "Non-Hodgkin's Lymphoma and Specific Pesticide Exposures in Men: Cross-Canada Study of Pesticides and Health." *Cancer Epidemiology, Biomarkers & Prevention* 10, no. 11 (November 2001): 1155–63, PubMed ID 11700263.

5   M. Eriksson, L. Hardell, M. Carlberg, and M. Akerman, "Pesticide Exposure as Risk Factor for Non-Hodgkin Lymphoma Including Histopathological Subgroup Analysis," *International Journal of Cancer* 123, no. 7 (October 2008): 1657–63, https://doi.org/10.1002/ijc.23589.

6   A. De Roos et al., "Integrative Assessment of Multiple Pesticides as Risk Factors for Non-Hodgkin's Lymphoma Among Men," *Occupational & Environmental Medicine* 60, no. 9 (September 2003): e11, https: //doi.org/10.1136/oem.60.9.e11.

Chapter 7: Daubert Day

1   Wallace Turner, "Rep. Phillip Burton, Democratic Liberal, Dies on Visit to California," *New York Times*, April 11, 1983, https://www.nytimes. com/1983/04/11/obituaries/rep-phillip-burton-democratic-liberal -dies-on-visit-to-california.html.

2   Christopher J. Portier, "A Comprehensive Analysis of the Animal Carcinogenicity Data for Glyphosate from Chronic Exposure Rodent Carcinogenicity Studies," *Environmental Health* 19, February 12, 2020, https://ehjournal.biomedcentral.com/articles/10.1186/s12940-020 -00574-1, accessed April 28, 2022.

Chapter 8: The Johnson Trial Begins

1   Lucia Fernandez, "Monsanto's Net Sales from 2008 to 2017," *Statista*, July 6, 2021, https://www.statista.com/statistics/276270/net-sales-and -net-income-of-monsanto-since-2008/#:~:text=In%202017%2C%20 Monsanto%20reported%20net,completed%20on%20June%20 7%2C%202018.

2   "George Lombardi Named 'Litigator of the Year' by The American Lawyer," Winston & Strawn LLP, January, 2014, https://www.winston. com/en/thought-leadership/george-lombardi-named-litigator-of -the-year-by-the-american.html; Amanda Bronstad, "In No. 3 Verdict,

Monsanto Scores Over Seed Patent," *The National Law Journal*, Special Report, March 4, 2013, accessed 03/24/2022, https://www.winston.com/images/content/5/9/v2/59070/100verdicts.pdf.

3 Michael Lefkowitz, "Illinois Appeals Court Reinstates $10 Billion 'Light' Cigarettes Verdict," LexisNexis, April 30, 2014, https://www.lexisnexis.com/legalnewsroom/litigation/b/litigation-blog/posts/illinois-appeals-court-reinstates-10-billion-light-cigarettes-verdict

4 Robin Mesnage, Charles Benbrook, and Michael N. Antoniou, "Insight into the Confusion Over Surfactant Co-Formulants in Glyphosate-Based Herbicides," *Food and Chemical Toxicology* 128 (June 2019): 137–45, https://www.sciencedirect.com/science/article/pii/S0278691519301814; Sarantis Michalopoulos, "EU Agrees Bans on Glyphosate Co-Formulant," EURACTIV, July 11, 2016, accessed 03/24/2022, https://www.euractiv.com/section/agriculture-food/news/eu-agrees-ban-on-glyphosate-co-formulant/.

5 C. Bolognesi, et al., "Biomonitoring of Genotoxic Risk in Agricultural Workers from Five Colombian Regions: Association to Occupational Exposure to Glyphosate," *Journal of Toxicology and Environmental Health,* Part A, 72, no. 15-16 (2009): 986-97; C. Bolognesi and N. Holland, "The Use of Lymphocyte Cytokinesis-Block Micronucleus Assay for Monitoring Pesticide-Exposed Populations," *Mutation Research*, Part A, 770 (October–December 2016): 183-203; E. Chang and E. Delzell, "Systematic Review and Meta-analysis of Glyphosate Exposure and Risk of Lymphohematopoietic Cancers," *Journal of Environmental Science and Health,* Part B, 51, no. 6 (2016):402–34; M. Kwiatkowska, et al., "DNA Damage and Methylation Induced by Glyphosate in Human Peripheral Blood Mononuclear Cells (In Vitro Study)," *Food and Chemical Toxicology* 105 (July 2017; epub March 27, 2017):93–8. doi: 10.1016/j.fct.2017.03.051; M. Kwiatkowska, B. Huras, and B. Bukowska, "The Effect of Metabolites and Impurities of Glyphosate on Human Erythrocytes (in Vitro)," *Pesticide Biochemistry and Physiology,* 109 (February 2014) : 34-43.

6 J. George, S. Prasad, Z. Mahmood, and Y. Shukla, "Studies on Glyphosate-Induced Carcinogenicity in Mouse Skin: A Proteomic Approach," *Journal of Proteome Research* 73, no. 5 (March 2010): 951–64.

7 Christopher J. Portier, "A Comprehensive Analysis of the Animal Carcinogenicity Data for Glyphosate from Chronic Exposure Rodent Carcinogenicity Studies," *Enviromental Health* 19, no. 18 (2020): 18, accessed March 24, 2022, https://ehjournal.biomedcentral.com/track/pdf/10.1186/s12940-020-00574-1.pdf. This is an excellent summary of glyphosate animal studies.

8 Christopher J. Portier et al. "Differences in the Carcinogenic Evaluation of Glyphosate between the International Agency for Research on Cancer

(IARC) and the European Food Safety Authority (EFSA)," *Journal of Epidemiology & Community Health* 70, no. 8 (2016): 741–45, accessed March 24, 2022, https://jech.bmj.com/content/70/8/741.long.

9   Carey Gillam and Nathan Donley, "A Story Behind the Monsanto Cancer Trial — Journal Sits on Retraction," *Environmental Health News*, August 22, 2018, accessed March 24, 2022, https://www.ehn.org /monsanto-science-ghostwriting--2597869694/no-action-has -been-taken.

10   Gary M. Williams et al., "A Review of the Carcinogenic Potential of Glyphosate by Four Independent Expert Panels and Comparison to the IARC Assessment," Supplement 1, *Critical Reviews in Toxicology* 46 (2016): 3–20, https://www.tandfonline.com/doi/full/10.1080/10408 444.2016.1214677; John Acquavella et al., "Glyphosate Epidemiology Expert Panel Review: A Weight of Evidence Systematic Review of the Relationship between Glyphosate Exposure and Non-Hodgkin's Lymphoma or Multiple Myeloma," Supplement 1,*Critical Reviews in Toxicology* 46 (2016): 28–43, https://www.tandfonline.com/doi/full/1 0.1080/10408444.2016.1214681?src=recsys; Gary M. Williams et al., "Glyphosate Rodent Carcinogenicity Bioassay Expert Panel Review," Supplement 1, *Critical Reviews in Toxicology* 46 (2016): 45–55, https:// www.tandfonline.com/doi/full/10.1080/10408444.2016.1214679?sr c=recsys; Helmut Greim et al., "Evaluation of Carcinogenic Potential of the Herbicide Glyphosate, Drawing on Tumor Incidence Data from Fourteen Chronic/Carcinogenicity Rodent Studies," *Critical Reviews in Toxicology* 45, no. 3 (2015): 185–208, https://www.tandfonline.com/ doi/full/10.3109/10408444.2014.1003423?src=recsys; Larry Kier and David J. Kirkland, "Review of Genotoxicity Studies of Glyphosate and Glyphosate-Based Formulations," *Critical Reviews in Toxicology* 43, no. 4 (2013): 283–315, https://www.tandfonline.com/doi/full/10.3109/1040 8444.2013.770820?src=recsys.

11   Gary M. Williams et al. "A Review of the Carcinogenic Potential of Glyphosate by Four Independent Expert Panels and Comparison to the IARC Assessment," Supplement 1, *Critical Reviews in Toxicology* 46 (2016), accessed March 24, 2022, https://www.tandfonline.com/doi/full /10.1080/10408444.2016.1214677.

12   Nathan Donley and Carey Gillam, "Public Records Sought on Monsanto's Ties to EPA's Pesticide Office," Center for Biological Diversity, April 14, 2017, https://www.biologicaldiversity.org/news /press_releases/2017/pesticides-04-14-2017.php accessed March 24, 2022.

13   All of the internal materials regarding this matter that were admitted as evidence in the trial can be viewed at https://www.baumhedlundlaw. com/documents/pdf/monsanto-documents-2/McClellan-Roger -Exhibit-05-Redacted-final.pdf.

14 "Expression of Concern - 26 September 2018," *Critical Reviews in Toxicology* 48, no. 10 (2018): 891, https://www.tandfonline.com/doi/full/10.1080/10408444.2018.1522786; "Expression of Concern - 30 November 2018," *Critical Reviews in Toxicology* 48, no. 10 (2018): 903, https://www.tandfonline.com/doi/full/10.1080/10408444.2018.1539570.

15 "R. Brent Wisner," Baum Hedlund Aristei & Goldman, https://www.baumhedlundlaw.com/attorneys/r-brent-wisner/.

16 Proposition 65 warning for glyphosate, see "Glyphosate," CA.gov, https://www.p65warnings.ca.gov/fact-sheets/glyphosate and https://www.p65warnings.ca.gov/chemicals/glyphosate.

Chapter 9: My Trial Testimony: Johnson v. Monsanto

1 "Bayer to Acquire Monsanto – What Does That Mean to Sustainability?" Futures Centre, November 7, 2017, https://www.thefuturescentre.org/signal/bayer-to-acquire-monsanto-what-does-this-mean-for-sustainability/, accessed March 25, 2022.

2 Ruth Bender, "How Bayer-Monsanto Became One of the Worst Corporate Deals—in 12 Charts," *The Wall Street Journal,* August 28, 2019, accessed March 25, 2022, https://www.wsj.com/articles/how-bayer-monsanto-became-one-of-the-worst-corporate-dealsin-12-charts-11567001577.

3 "Bayer Closes Monsanto Acquisition," Bayer, June 7, 2018, https://media.bayer.com/baynews/baynews.nsf/id/Bayer-closes-Monsanto-acquisition.

4 Robert Teitelman, "Bayer's Deal for Monsanto Looked Like a Winner. Now It Looks Like a Lesson in How Not to Do M&A," *Barron's,* March 22, 2019, https://www.barrons.com/articles/bayers-acquisition-of-monsanto-how-not-to-do-m-a-51553296373.

5 Bender, "How Bayer-Monsanto Became One of the Worst Corporate Deals—in 12 Charts."

6 Andrew Noël, "How Monsanto's Roundup Herbicide Went from Bayer Asset to Burden," *Bloomberg Law,* June 24, 2020, https://news.bloomberglaw.com/environment-and-energy/how-monsantos-roundup-herbicide-went-from-bayer-asset-to-burden.

7 Noël, "How Monsanto's Roundup Herbicide Went from Bayer Asset to Burden."

8 Steven H. Swerdlow et al., "The 2016 Revision of the World Health Organization Classification of Lymphoid Neoplasms," *Blood* 127, no. 20 (2016): 2375–90, https://ashpublications.org/blood/article/127/20/2375/35286/The-2016-revision-of-the-World-Health-Organization.

9   A. J. De Roos, et al., "Integrative Assessment of Multiple Pesticides as Risk Factors for Non-Hodgkin's Lymphoma among Men," *Occupational & Environmental Medicine* 60, no. 9 (2003): e11, https://oem.bmj.com /content/60/9/e11.

Chapter 10: The Johnson Verdict

1   "judgment notwithstanding the verdict (JNOV)," Legal Information Institute, last updated June 2020, https://www.law.cornell.edu/wex /judgment_notwithstanding_the_verdict_(jnov)#:~:text=Primary%20 tabs,the%20timing%20within%20a%20trial.

Chapter 11: Hardeman

1   See 28 U.S.C. § 1441. — from https://www.law.cornell.edu/uscode /text/28/1441#:~:text=Except%20as%20otherwise%20expressly%20 provided,district%20and%20division%20embracing%20the.
2   Sam Levin, "The Family That Took on Monsanto: 'They Should've Been with Us in the Chemo Ward,'" *The Guardian*, April 10, 2019, https://www.theguardian.com/business/2019/apr/10/edwin-hardeman -monsanto-trial-interview.
3   Levin, "Family That Took on Monsanto."
4   See Hyuna Sung et al., "Emerging Cancer Trends Among Young Adults in the USA: Analysis of Population-Based Cancer Registry, *Lancet Public Health* 4, no. 3 (March 2019): E137–47, https://doi.org/10.1016/S2468- 2667(18)30267-6 (see also Supplementary Appendix in that volume of the *Lancet*); Eleanor V. Willett et al., "Non-Hodgkin Lymphoma and Obesity: A Pooled Analysis from the InterLymph Consortium," *International Journal of Cancer* 122, no. 9 (May 2008): 2062–70, https:// doi.org/10.1002/ijc.23344; Lindsay M. Morton et al., "Etiologic Heterogeneity Among Non-Hodgkin Lymphoma Subtypes: The InterLymph Non-Hodgkin Lymphoma Subtypes Project," *Journal of the National Cancer Institute, Monographs* 2014, no. 48 (August 2014): 130–44, https://doi.org/10.1093/jncimonographs/lgu013; Jorge J. Castillo et al., "Obesity Is Associated with Increased Relative Risk of Diffuse Large B-Cell Lymphoma: A Meta-Analysis of Observational Studies," *Clinical Lymphoma, Myeloma, & Leukemia* 14, no. 2 (April 2014): 122–30, https://doi.org/10.1016/j.clml.2013.10.005; Brian C-H Chiu et al., "Obesity and Risk of Non-Hodgkin Lymphoma (United States)," *Cancer Causes & Control* 18, no. 6 (August 2007): 677–85, https://doi .org/10.1007/s10552-007-9013-9; James R. Cerhan et al., "Medical

History, Lifestyle, Family History, and Occupational Risk Factors for Diffuse Large B-Cell Lymphoma: The InterLymph Non-Hodgkin Lymphoma Subtypes Project," *Journal of the National Cancer Institute, Monographs* 2014, no. 48 (August 2014): 15–25, https://doi.org/10.1093/jncimonographs/lgu010.

5   Select references pertaining to Hepatitis C: Y. Kawamura et al., "Viral Elimination Reduces Incidence of Malignant Lymphoma in Patients with Hepatitis C.," *American Journal of Medicine* 120, no. 12 (2007): 1034–41; A. Pellicelli, et al., "Antiviral Therapy in Hepatitis C-Infected Patients Prevents Relapse of Diffuse Large B Cell Lymphoma," *Journal of Clinical and Experimental Hepatology* 4, no. 3 (2018): 197–200; Y. Tsutsumi Y, et al., "Efficacy and Prognosis of Antiviral Therapy on Hepatitis C Following Treatment of Lymphoma in HCV-Positive Diffuse Larce-Cell Lymphoma," *Annals of Hematology* 96, no. 12 (2017):2057–61.

Chapter 12:   The Second Trial Begins

1   Sam Levin, "Monsanto: Judge Threatens To Shut Down Cancer Patient's Lawyer," *The Guardian*, February 25, 2019, https://www.theguardian.com/business/2019/feb/25/monsanto-federal-trial-roundup-cancer.

2   One of the papers she shared with the jury that addresses this topic is by Andreas Stang, Charles Poole, and Oliver Kuss, "The Ongoing Tyranny of Statistical Significance Testing in Biomedical Research," *European Journal of Epidemiology* 25, no. 4 (April 2010): 225–230.

3   John F. Acquavella, Bruce H. Alexander, Jack S. Mandel, Carol J. Burns, and Christophe Gustin, "Exposure Misclassification in Studies of Agricultural Pesticides: Insights from Biomonitoring," *Epidemiology* 17, no. 1 (January 2006): 69–74, https://doi.org/10.1097/01.ede.0000190603.52867.22.

4   George M. Gray et al., "The Federal Government's Agricultural Health Study: A Critical Review with Suggested Improvements," *Human and Ecological Risk Assessment: An International Journal* 6, no. 1 (2000): 47–71, https://doi.org/10.1080/10807030091124446.

5   Scott Weichenthal, Connie Moase, and Peter Chan, "A Review of Pesticide Exposure and Cancer Incidence in the Agricultural Health Study Cohort," *Environmental Health Perspectives* 118, no. 8 (August 2010): 1117–25, https://doi.org/10.1289/ehp.0901731; Aaron Blair et al., "Impact of Pesticide Exposure Misclassification on Estimates of Relative Risks in the Agricultural Health Study," *Occupational & Environmental Medicine* 68, no. 7 (July 2011): 537–41, https://doi.org/10.1136/oem.2010.059469.

6 Hans-Olov Adami, David J. Hunter, Pagona Lagiou, and Lorelei Mucci, eds., *Textbook of Cancer Epidemiology*, 3rd ed. (New York: Oxford University Press, 2018).

Chapter 13: Verdicts

1 Bayer, "Bayer Statement on Jury's Decision in Phase One of California Glyphosate Trial," news release, March 19, 2019, website last updated August 1, 2021, https://media.bayer.com/baynews/baynews.nsf /id/Bayer-statement-on-jurys-decision-in-phase-one-of-California -glyphosate-trial.
2 Email from Donna Farmer to Sekhar Nataranjan, "Subject: Agitation against Roundup," November 24, 2003, *Hardeman v. Monsanto* documents, https://www.baumhedlundlaw.com/documents/pdf /monsanto-documents/27-Internal-Monsanto-Email-You-Cannot-Say -That-Roundup-is-not-a-Carcinogen.pdf.
3 "$80 Million Awarded to Man Who Jury Says Got Cancer from Roundup Exposure," CBS 13 Sacramento, March 27, 2019, https://sacramento. cbslocal.com/2019/03/27/80-million-awarded-to-man-who-jury -says-got-cancer-from-roundup-exposure/.

Chapter 14: The Pilliods

1 Paolo Bottetta and Frank de Vocht, "Occupation and the Risk of Non-Hodgkin Lymphoma," *Cancer Epidemiology, Biomarkers & Prevention* 16, no. 3 (March 2007): 369–72, https://doi.org/10.1158/1055-9965 .EPI-06-1055.
2 Leah Schinasi and Maria E. Leon, "Non-Hodgkin Lymphoma and Occupational Exposure to Agricultural Pesticide Chemical Groups and Active Ingredients: A Systematic Review and Meta-Analysis, *International Journal of Environmental Research and Public Health* 11 no. 4 (April 2014): 4449–527, https://doi.org/10.3390/ijerph110404449.
3 The specific statements are as follows: "Our meta-analysis yielded borderline significant RRs of 1.3 and 1.4 between glyphosate use and risk of NHL and MM" (423) and "We found marginally significant positive meta RRs for the association between glyphosate use and risk of NHL and MM" (424). Ellen Chang and Elizabeth Delzell, "Systematic Review and Meta-Analysis of Glyphosate Exposure and Risk of Lymphohematopoietic Cancers," *Journal of Environmental Science and Health,* Part B, 51, no. 6 (2016): 402–34, https://doi.org/10.1080/0360123 4.2016.1142748.

4 Luoping Zhang, Iemaan Rana, Rachel M. Shaffer, Emanuela Taioli, and Lianne Sheppard, "Exposure to Glyphosate-Based Herbicides and Risk for Non-Hodgkin Lymphoma: A Meta-Analysis and Supporting Evidence," *Mutation Research. Reviews in Mutation Research* 781 (July–September 2019): 186–206, https://doi.org/10.1016/j.mrrev.2019.02.001.

Chapter 15: The Third Trial Begins

1 For further explanation, see International Agency for Research on Cancer, "Questions and Answers," IARC Monographs on the Identification of Carcinogenic Hazards to Humans, World Health Organization, December 20, 2019, https://www.iarc.who.int; https://monographs.iarc.who.int/wp-content/uploads/2019/07/Preamble-2019.pdf; International Agency for Research on Cancer, "Glyphosate," June 2018, https://monographs.iarc.who.int/wp-content/uploads/2018/06/mono112-10.pdf.
2 M. Fallah et al., "Autoimmune Diseases Associated with Non-Hodgkin Lymphoma: A Nationwide Cohort Study," *Annals of Oncology* 25, no. 10 (October 2014): 2025–30, https://doi.org/10.1093/annonc/mdu365.
3 Brian C-H Chiu, et al., "Agricultural Pesticide Use and Risk of t(14;18) -Defined Subtypes of Non-Hodgkin Lymphoma," [Blood] 108, no.4 (August 2006): 1363–69, https://doi.org/10.1182/blood-2005-12-008755.
4 "Emails entered into evidence, available at https://www.baumhedlundlaw.com/documents/pdf/monsanto-documents-2/Monsanto-execs-playing-whack-a-mole.pdf.
5 Charles M. Benbrook, "How Did the US EPA and IARC Reach Diametrically Opposed Conclusions on the Genotoxicity of Glyphosate -Based Herbicides?" *Environmental Sciences Europe* 31, no. 2 (2019), https://enveurope.springeropen.com/articles/10.1186/s12302-018-0184-7.

Chapter 16: Another Day in Court: The Pilliods vs. Monsanto

1 Sylvain Lamure, Camille Carles, Quam Aquereburu, et al., "Association of Occupation Pesticide Exposure with Immunochemotherapy Response and Survival among Patients with Diffuse Large B-Cell Lymphoma," *JAMA Network Open* 2, no. 4 (2019): e192093, https://jamanetwork.com/journals/jamanetworkopen/fullarticle/2730779.

Chapter 17: The $2 Billion Judgment

1 Andrew Noël, "How Monsanto's Roundup Herbicide Went from Bayer Asset to Burden," *Bloomberg Law*, June 24, 2020, https://news .bloomberglaw.com/environment-and-energy/how-monsantos -roundup-herbicide-went-from-bayer-asset-to-burden.

2 Ruth Bender, "Bayer Shareholders Signal Loss of Confidence in CEO," *Wall Street Journal*, April 26, 2019, https://www.wsj.com/articles /bayer-ceo-faces-shareholder-ire-over-monsanto-deal-11556292088.

3 "Annual Stockholders' Meeting: Address by Werner Baumann," Bayer press release, April 26, 2019, https://media.bayer.com/baynews /baynews.nsf/id/BBLB9P-Address-by-Werner-Baumann.

4 EPA, "DDT—A Brief History and Status," accessed March 27, 2022, https://www.epa.gov/ingredients-used-pesticide-products/ddt-brief -history-and-status.

5 Emily Holden, "Trump EPA Insists Monsanto's Roundup Is Safe, despite Cancer Cases," *Guardian*, April 30, 2019, https://www .theguardian.com/business/2019/apr/30/monsanto-roundup-trump -epa-cancer.

6 "EPA Takes Next Step in Review Process for Herbicide Glyphosate, Reaffirms No Risk to Public Health," EPA press release, April 30, 2019, https://www.epa.gov/newsreleases/epa-takes-next-step-review -process-herbicide-glyphosate-reaffirms-no-risk-public.

7 Patricia Cohen, "$2 Billion Verdict against Monsanto Is Third to Find Roundup Caused Cancer," *New York Times*, May 13, 2019, https://www .nytimes.com/2019/05/13/business/monsanto-roundup-cancer-verdict .html.

8 "Bayer Statement on Jury's Decision in California State Glyphosate Trial," Bayer press release, May 13, 2019, https://media.bayer.com /baynews/baynews.nsf/id/Bayer-statement-on-jurys-decision-in -California-State-Glyphosate-Trial.

Chapter 18: Endings

1 Antlee, "Not Your Time," YouTube, posted August 19, 2019, https:// youtu.be/EWr1Y61WtTY.

2 Greg Stohr, "Bayer Bid to End Roundup Suits Draws U.S. High Court Inquiry," Bloomberg Law, December 13, 2021, https://news. bloomberglaw.com/us-law-week/bayer-bid-to-end-roundup-suits -draws-u-s-supreme-court-inquiry.

3 Hannah Albarazi, "Calif. Judge Rips Monsanto's Atty in $87M Roundup Appeal," Law360, June 22, 2021, https://www.law360.com/articles /1396471/calif-judge-rips-monsanto-s-atty-in-87m-roundup-appeal.

4 Matthew Renda, "Bayer to Pull Glyphosate from Stores Due to Cancer Lawsuits," Courthouse News Service, July 29, 2021, https://www .courthousenews.com/bayer-to-pull-glyphosate-from-stores-due-to -cancer-lawsuits/; Alex Robinson, "Bayer to Stop Residential Sales of Glyphosate-Based Roundup by 2023," Modern Farmer, July 29, 2021, https://modernfarmer.com/2021/07/bayer-to-stop-residential -sales-of-roundup-by-2023/.

5 "Bayer Provides Update on Path to Closure of Roundup Litigation," Bayer press release, July 29, 2021, https://media.bayer.com/baynews /baynews.nsf/id/Bayer-Provides-Update-on-Path-to-Closure -of-Roundup-Litigation; "Five-Point Plan to Close the Roundup Litigation," Bayer press release, August 16, 2021, https://www.bayer .com/en/roundup-litigation-five-point-plan#:~:text=To%20further%20 reduce%20future%20litigation,because%20of%20any%20safety%20 concerns.

ADDITIONAL RESOURCES

Roundup and glyphosate

California Office of Environmental Health Hazard Assessment. "Glyphosate." Chemicals Considered or Listed Under Proposition 65. Accessed March 18, 2022. https://oehha.ca.gov/proposition-65/chemicals/glyphosate.

Duke, Stephen O., and Stephen B. Powles. "Mini-Review: Glyphosate: A Once-in-a-Century Herbicide." *Pest Management Science* 64, no. 4 (2008): 319–25. https://doi.org/10.1002/ps.1518.

Monnier, Jen. "What Is Glyphosate?" LiveScience. Updated September 21, 2020. https://www.livescience.com/glyphosate-round-up.html.

National Pesticide Information Center. "Glyphosate: General Fact Sheet." Last modified March 2019. http://npic.orst.edu/factsheets/glyphogen.html.

Smith-Schoenwalder, Cecelia. "What to Know about Glyphosate, the Pesticide in Roundup Weed Killer." *U.S. News and World Report.* August 19, 2019. https://www.usnews.com/news/national-news/articles/what-to-know-about-glyphosate-the-pesticide-in-roundup-weed-killer.

Turner, Terry. "Roundup." ConsumerNotice.org. Last modified October 7, 2021. http://www.consumernotice.org/environmental/pesticides/roundup.

The IARC

"Agents Classified by the IARC Monographs, Volumes 1–130." IARC Monographs on the Identification of Carcinogenic Hazards to Humans, International Agency for Research on Cancer, World Health Organization. Last updated April 8, 2022. https://monographs.iarc.fr/agents-classified-by-the-iarc.

Alleged ghostwriting

Donley, Nathan, Bill Freese, Emily Marquez, and Caroline Cox. "Letter to the Editors of *Critical Reviews in Toxicology*." n.d. Center for Biological

Diversity website. Accessed April 10, 2022. https://www
.biologicaldiversity.org/campaigns/pesticides_reduction/pdfs
/Retraction_letter_to_Critical_Reviews_in_Toxicology.pdf.
Krimsky, Sheldon. "Monsanto's Ghostwriting and Strong-Arming Threaten
Sound Science—and Society." Environmental Health News. June 26,
2018. https://www.ehn.org/monsanto-effort-to-skew
-science-2581194459.html.

Daubert hearings

Malkan, Stacy. "Report from the Glyphosate Daubert Hearings." U.S. Right
to Know. Last updated March 19, 2018. https://usrtk.org/pesticides
/reports-from-cancer- victims-v-monsanto-daubert-hearings.

Johnson trial

Dewayne Johnson v. Monsanto Company, et al. "Index of Proceedings."
Case no. CGC-16-550128. Superior Court of California. July 12, 2018.
www.baumhedlundlaw.com/pdf/monsanto-documents/johnson-trial
/Johnson-Day-Three-A-7-12-18.pdf.

Hardeman trial

Edwin Hardeman v. Monsanto Company. "Opinion." United States
Court of Appeals for the Ninth Circuit. https://usrtk.org/wp
-content/uploads/2021/05/9th-circuit-appeals-court-ruling-on
-Hardeman-v-Monsanto.pdf.
Hardeman v. Monsanto trial transcript. February 4, 2019. https://usrtk
.org/wp-content/uploads/bsk-pdf-manager/2019/02/Transcript
-of-Feb.-4-2019-hearing-in-front-of-judge-Chhabria.pdf.
Monsanto v. Edwin Hardeman. "Petition for a Writ of Certiorari." U.S.
Supreme Court. https://usrtk.org/wp-content/uploads/2021/08
/Bayer-petitions-to-Sup-Crt.pdf.
United States Court of Appeals for the Ninth Circuit. "19-16253 Edwin
Hardeman v. Monsanto Company." YouTube, posted October 24, 2020.
https://www.youtube.com/watch?v=OzX8vz5LoMo.
United States District Court, Northern District of California.
"RE: Roundup Products Liability Litigation." Document 2791. Filed
February 24, 2019. https://www.cand.uscourts.gov/filelibrary/3606
/PTO85.pdf. https://soygrowers.com/wp-content/uploads/2022
/05/5-23-22-Glyphosate-Solicitor-General-Letter-1.pdf.
https://www.reuters.com/legal/government/us-supreme-court
-rejects-bayer-bid-nix-roundup-weedkiller-suits-2022-06-21/.

Pilliods trial

Alva Pilliod, et al. v. Monsanto Company. "Plaintiff's Notice of Acceptance of Remittitur." Case No. RG17862702. Superior Court of the State of California for the County of Alameda. https://www.baumhedlundlaw.com/documents/pdf/monsanto-documents/pilliod/Plaintiffs-Notice-of-Acceptance-of_Remittitur-20190726121214.pdf. https://www.bloomberg.com/news/articles/2022-06-27/bayer-loses-again-as-high-court-allows-87-million-roundup-award. https://www.reuters.com/business/us-supreme-court-again-nixes-bayer-challenge-weedkiller-suits-2022-06-27/.

After the trials

Burger, Ludwig, and Patricia Weiss. "Bayer Takes $10 Billion Writedown, Flags Higher Roundup Settlement Bill." Reuters. November 3, 2020. https://www.reuters.com/article/bayer-results/bayer-takes-10-billion-writedown-flags-higher-roundup-settlement-bill-idUSKBN27J26A.

Gillam, Carey. "Bayer Seeks U.S. Supreme Court Review of Roundup Trial Loss." U.S. Right to Know. August 16, 2021. https://usrtk.org/monsanto-roundup-trial-tracker/bayer-seeks-u-s-supreme-court-review-of-roundup-trial-loss/.

Lavoie, Denise. "Virginia Attorney Charged in Extortion Plot over Roundup." Federal News Network. December 19, 2019. https://federalnewsnetwork.com/business-news/2019/12/virginia-attorney-charged-in-extortion-plot-over-roundup/.

Thomas, David. "Virginia Lawyers Get Prison Terms for $200M Roundup Extortion Scheme." Reuters. September 18, 2020. https://www.reuters.com/article/lawyers-sentencing/virginia-lawyers-get-prison-terms-for-200m-roundup-extortion-scheme-idUSL1N2GF2FH.

United States District Court, Northern District of California. "In Re: Roundup Products Liability Litigation." Document 12531. https://usrtk.org/wp-content/uploads/2021/05/Judge-denies-Bayer-class-plan.pdf. https://www.reuters.com/world/us/biden-administration-asks-us-supreme-court-shun-bayer-weedkiller-appeal-2022-05-10/ https://www.supremecourt.gov/DocketPDF/21/21-241/222984/20220510154200610_Monsanto.CVSG%205.9.22%20v.2.pdf. https://www.justice.gov/opa/pr/virginia-attorneys-plead-guilty-orchestrating-200-million-extortion-scheme-targeting.

INDEX

Page numbers in *italics* refer to figures.